Alexandr I. Kozlow

The fundamentals of ultra-high dimensional mosaic FPs

Alexandr I. Kozlow

The fundamentals of ultra-high dimensional mosaic FPs

ScienciaScripts

Imprint
Any brand names and product names mentioned in this book are subject to trademark, brand or patent protection and are trademarks or registered trademarks of their respective holders. The use of brand names, product names, common names, trade names, product descriptions etc. even without a particular marking in this work is in no way to be construed to mean that such names may be regarded as unrestricted in respect of trademark and brand protection legislation and could thus be used by anyone.

Cover image: www.ingimage.com

This book is a translation from the original published under ISBN 978-620-3-40956-7.

Publisher:
Sciencia Scripts
is a trademark of
International Book Market Service Ltd., member of OmniScriptum Publishing Group
17 Meldrum Street, Beau Bassin 71504, Mauritius
Printed at: see last page
ISBN: 978-620-3-31995-8

Copyright © Alexandr I. Kozlow
Copyright © 2021 International Book Market Service Ltd., member of OmniScriptum Publishing Group

RUSSIAN SCIENTISTS

TODAY:

Anatoly Vasilievich Rzhanov,

Konstantin Konstantinovich Svitashev,

Boris Ivanovich Fomin,

Valery Vladimirovich Shashkin,

Anatoly Grigorievich Klimenko

Alexander I. Kozlov

FUNDAMENTALS OF CREATING SUPER-HIGH-DIMENSIONAL MOSAIC PHOTODETECTORS WITH MAXIMUM IMAGE CONVERSION EFFICIENCY

ANALYTICAL REVIEW: STUDY OF THE ULTIMATE NANO- AND MICROMINIATURIZATION OF ULTRA-HIGH DIMENSIONAL MOSAIC PHOTODETECTORS

Novosibirsk

2021

UDC 621.382 : 535.231.62

LBC 32.854.12

K 59

Kozlov A. I. Fundamental bases of creating mosaic photodetectors of ultrahigh dimensionality with maximum image conversion efficiency. Analytical review. Ed. 3, revised. - Latvia, Riga: Lambert Academic Publishing. 2021. - 82 p., ill.

ISBN 978-620-3-40956-7

In the analytical review the mosaic technology is investigated as a promising fundamental basis for the ultimate nano- and micro-miniaturization of photodetectors with achieving high and ultra-high dimensionality. The most characteristic and interesting fundamental approaches and models, theoretical and semiempirical prototypes, developed technological layouts and created research samples of mosaic photodetectors (MFDs) of ultra-high dimensionality with maximum image conversion efficiency are analyzed in the comparison. A methodology for comparative analysis of image conversion efficiency in MFPs is developed. The principle of operation of MFPs of extra high dimensionality is presented. The modern fundamental groundwork for the creation of MFPs with extremely high image conversion efficiency has been formed. Technological prototypes of MFPs of extra high dimensionality have been proposed. An optical methodology for achieving the ultimate efficiency of image conversion in the creation of MFPs has been developed. It is shown that it is possible to create MFPs without losses of elements in the formed images. The methodology for analyzing the temperature resolution of MFPs was further investigated. The achieved level of mosaic technology of high and ultrahigh dimensional photodetectors with maximum image conversion efficiency is considered.

For scientists and technicians engaged in research, development, and application of high and ultra-high dimensional photodetectors for various spectral ranges.

Table 2. Fig. 39. Bibliography. 109 titles.

ISBN 978-620-3-40956-7

5

PREDICTION

In the 1978 Geological Dictionary, the term "mosaic crystal" denotes a crystal with a structure of separate homogeneous blocks, somewhat displaced relative to each other and often separated from each other by internal gaps. However, we and our colleagues in nano- and microelectronics use the term "mosaic" to refer to multisubmodular devices (devices, emitters, photodetectors). In electronics, nano- and microelectronics, an informative analytical review of the term "mosaic" was conducted by the author in the period from 1983 to the present. Sometimes, the authors of other works use the adjective "composite"; however, from our point of view, the term "mosaic" most accurately defines the essence of the research object. The depth of the literary search is before the beginning of the mention of the term "mosaic", practically up to 1880. This analytical review presents the most characteristic and interesting fundamental approaches and models, theoretical and semi-empirical prototypes, developed technological layouts and created experimental samples of mosaic photodetectors of ultra-high dimensionality with maximum image conversion efficiency. The development of a methodology for analyzing the temperature resolution of MFPs is shown, and NETD estimates for MFPs based on multilayer structures with quantum wells are given. An optical methodology for achieving the ultimate image conversion efficiency in ultrahigh-dimensional MFP technology is further investigated. A comparative analysis of the experimental data shows the possibility of creating super high-dimensional   MHPs without loss of elements in the formed images. The results of the study are applicable to the manufacture of MHPs of high and extra-high dimensions, consisting of photodetector submodules based on uncooled microbolometers of infrared and terahertz ranges, multilayer structures with quantum wells and other detectors with different wavelength maximum spectral characteristics of photosensitivity.The monograph is intended for researchers and technical specialists engaged in research, development and application of high and extra-high dimension photodetectors.

Associate Professor, PhD in Physics and Mathematics A.G. Kharlamov, PhD in Engineering A.I. Kozlov, Professor, PhD in Physics and Mathematics V.N. Ovsyuk

FUNDAMENTALS OF THE CREATION OF ULTRA-HIGH-DIMENSIONAL MOSAIC PHOTODETECTORS
WITH MAXIMUM IMAGE CONVERSION EFFICIENCY

A. I. Kozlov1, Ph.

INTRODUCTION

A direct increase in the number of photosensitive elements (PSEs) in photodetectors (PDs) leads to a significant increase in the area of submodule crystals, which decreases the yield of good crystals, which directly increases the cost of PDs. However, the need to increase the spatial resolution of thermal imaging systems stimulates the creation of super-high-dimensional (UHD and more) FPs for the infrared (IR) and terahertz (THz) spectral ranges.

Mosaic technology is one of the promising, breakthrough technical solutions to radically increase the format of photodetectors. Mosaic photodetectors (MPS) are created by means of the corresponding technology on the basis of precision microassembly of butt to butt sub-module crystals of an acceptable format for manufacturing.

The main problem of MFPs is "blind zones" - areas along the junction lines of adjacent submodule crystals in which PFCs are absent. "Blind zones" are defined by the damage zones of semiconductor material and multilayer nano- and microstructures at the edges of MFP submodule crystals [1-54]. The size of the technological "blind zone" of MFP depends on the size of the damage areas of semiconductor materials and multilayer nano- and microstructures, the areas of roughness of the mating sides

---

[1] *Ph.D. Kozlov Alexander Ivanovich; e-mail: aikozlov13@mail.ru*

crystals and gaps between crystals of adjacent submodules; possible damage zones are shown in Figure 1 [1-16].

When projecting the input image onto the merged matrix, some elements are lost in the "blind spots" and, as a result, the efficiency2 (ηeff) of image conversion in MFP is reduced [1-10]. The size of the blind spots, in many respects, limits the applications of MFPs.

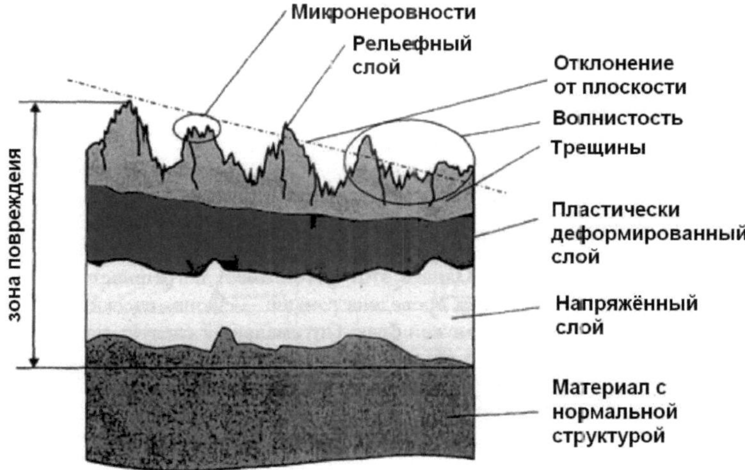

Figure 1 - Example of the structure of disturbances in semiconductor materials after cutting them with diamond discs [14, 15].

Reducing the number of elements in the "blind spots" of the MFP increases the amount of raw data for the video processor of readout photo signals. В

---

2 *Image conversion efficiency is the ratio of the number of input image elements received at the MFP optical outputs to the total number of input image elements focused on the combined MFP photodetector array [1-10].*

The result is increased reliability of subsequent processing and the quality of the imaged images [1-10]. Manufacturers of MFPs strive to reduce the size of blind spots.

The literature describes the following characteristic technological configurations of MFPs. First, MFPs consisting of several separate FPs - submodules (Figure 2a). Figure 2a shows a photo of a 4096×4096 MFP consisting of four FP-submodules [11]. Figure 2b shows a photo of FP-submodule with format 2048×2048, manufactured by "Raytheon Vision Systems" (USA). A variant of MFP of 4096×1024 dimension on the basis of four submodules of 1024×1024 format, arranged in a line, is shown in picture 3 [23]. Secondly, MFPs, consisting of one crystal of the PFC matrix and several multiplexer crystals. Multiplexer is an integral circuit for reading signals of the PFC matrix of MFP submodules. Figure 4 shows a photo of MFP by "ATLAS" (USA) on a common switching substrate. MFP consists of one crystal of PFC matrix and

16 multiplexers [24]. Third, one multiplexer crystal and several crystals of PFE matrices. In such MFPs the influence of different thermal expansion coefficients of multiplexer crystals and PFC matrices can be reduced [25].

For linear MFPs a technological solution was developed with the displacement of the edge elements up and down relative to the main PFC line (Figures 5-6) [26]. The photographic signals from the displaced edge PFCs are placed in the corresponding image locations in ROM and there are no "blind spots" in the formed image. However, the technological approaches used in the hybrid MFP based on linear PFC crystals are applicable, for the most part, just for linear submodules [26, 27].

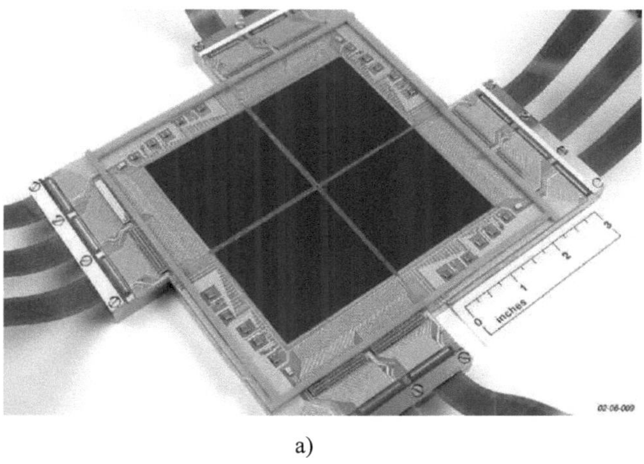

a)

Figure 2 *a* - Photograph of an MFP of 4096×4096 dimension (a), consisting of 4 FPs - 2048×2048 submodules (b) [11].

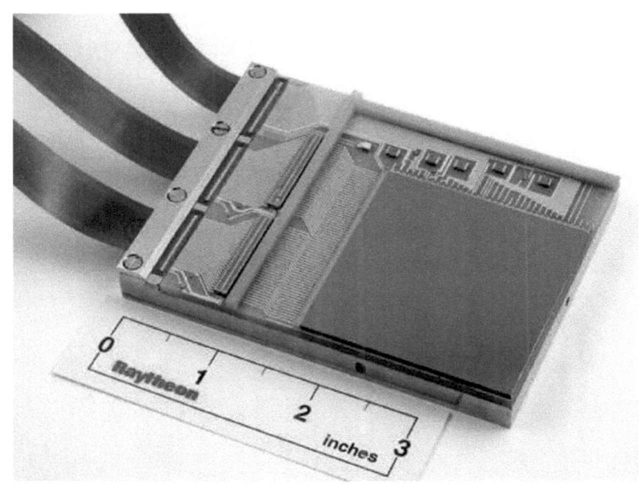

б)

Figure *2b* - Photo of the 2048×2048 format FP submodule by Raytheon Vision Systems (USA) [11].

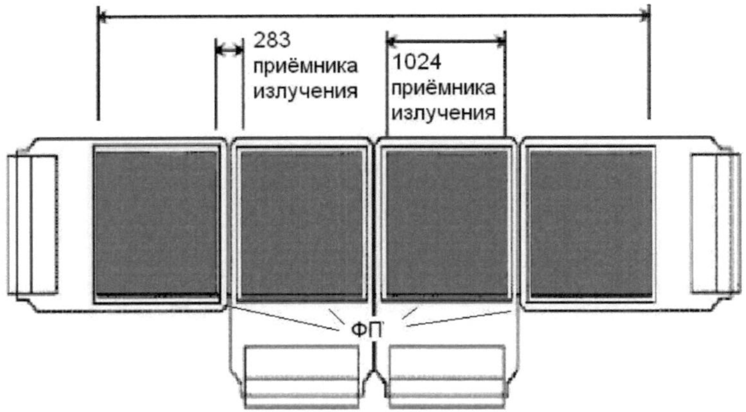

Figure 3 - MFP consisting of four 1024 × 1024 × 1024 × 1024 format FFEs with 27 μm FFE spacing; the "blind zone" is 383 FFE [23].

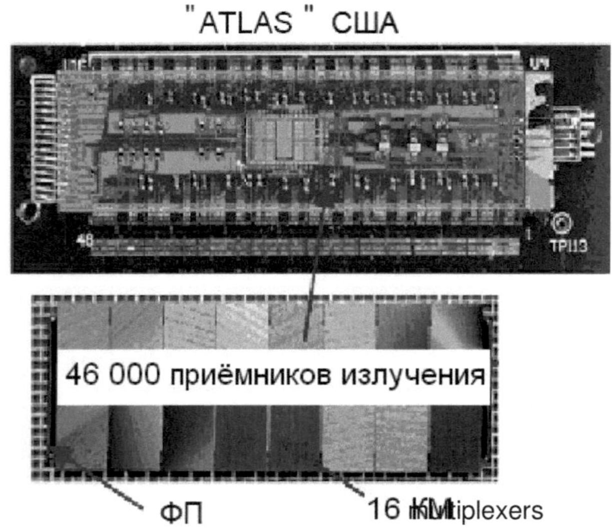

Figure 4 - Matrix MFP consisting of a single FFE matrix crystal and The multiplexer crystals, on a switching substrate [24].

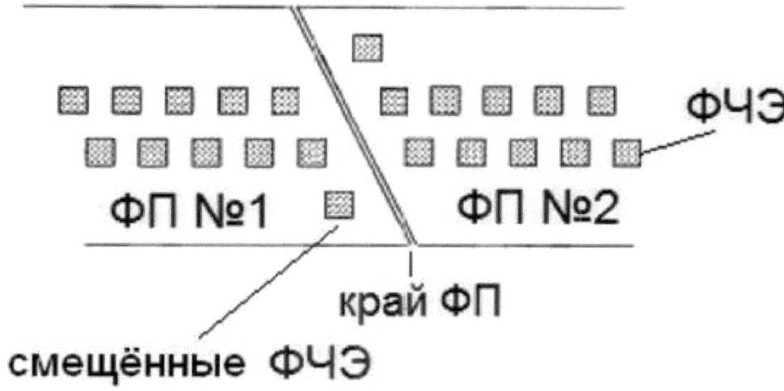

Figure 5 - Illustration of the method of manufacturing a linear MFP without loss of video signals of edge FFE submodules (firm "LETI", France) [26].

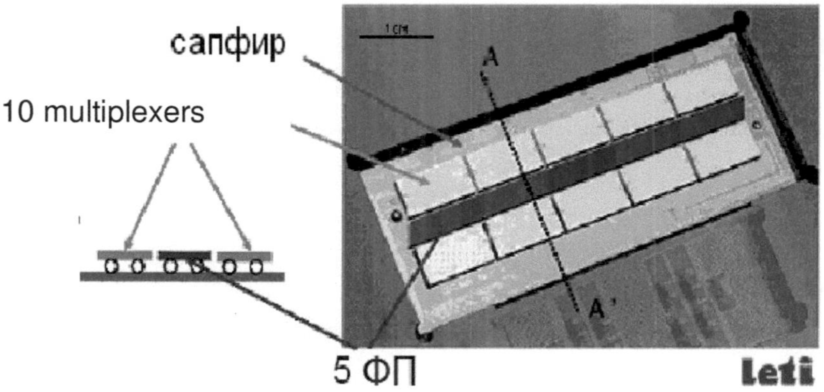

Figure 6 - Photograph of an MFP consisting of five 300×2 matrix crystals with an FFE step equal to 30 μm and 10 multiplexers [26].

In 2010, the size of the technological "blind spot" was reached $\geq$46 μm [1, 9, 10]. Since 2013, research and optimization of a prototype technology for creating MFPs with a total size of the "blind zone" between the edge FFEs of adjacent crystals of about 30 μm, including areas of rough crystal faces and the gap between crystals of about 20 μm [1, 2, 11-30]. Since 2014, one of the leading Russian firms expects to increase the size of the gap area to 10-20 μm [25].

In the technology of known MFPs, crystals are clamped from above using micromanipulators, which is unacceptable in the case of matrix microbolometer receivers (MMBRs), as well as other monolithic MFPs; such technological operation is not always possible for hybrid submodules [29, 37, 47]. Due to significant "blind spots", significant cost and large size of known MFPs, there are limitations of wide industrial application of MFPs for a number of sectors of the world economy.

The purpose of this paper is to provide an analytical review of ultrahigh intensity MFPs dimensionality and development of the fundamentals of mosaic technology with the achievement of the minimum size of "blind zones" between the edge PFCs of adjacent submodule crystals and, accordingly, the maximum efficiency of image conversion in MFPs.

In fundamental technological approaches to creation of MFPs and in nano- and microelectronic designs it is proposed to use MFPs as (cooled or uncooled) submodules: in monolithic version - frameless monolithic FPs in the form of multiplexer crystals with additionally integrated PFC arrays on them, and in hybrid version - frameless micro assemblies of matrix crystals of PFC and multiplexers. The ultimate miniaturization of super-high-dimensional MFPs with maximum image conversion efficiency is achieved precisely by applying frameless submodules with minimal (5 - 8 μm) damage regions of semiconductor materials and multilayer nano- and microstructures on the docked crystal edges and by placing frameless submodules on a single carrier plate with minimal (no more than 2-3 μm) gaps between crystals [1, 9, 10].

1. MAIN RESULTS OF THE LITERATURE REVIEW

1.1. TECHNOLOGICAL APPROACHES TO THE CREATION OF MFP

In [48] an 8192×4096 MFP, based on a matrix of 4×2 hybrid submodules of 2048×2048 elements (Figure 7), is considered. Submodule crystals are placed in open packages. An open-cased submodule is a hybrid assembly of two crystals. Crystals of the PFC matrix and multiplexer are connected using indium microtubes. As a part of MFP submodules in open packages are aligned in all coordinates using mechanical elements and fixed in a given position on the carrier.

However, this MFP has disadvantages (see Fig. 7): MFP consists of submodules with crystals of PFC matrices of the same spectral range; further increase in dimensionality is possible only by increasing the linear size of the submodule crystals used; alignment of submodules is performed using mechanical fastening elements, which determines large "blind spots" of 2.88 mm, which is equivalent to 160 PFC with 18 μm pitch.

In [26-28], a hybrid MFP based on linear FFE crystals was considered, in which a blind zone size of about 46 μm was achieved (see Fig. 5-6). This technological approach is applicable to linear submodules [27].

In [25] a photodetector device (PDD) containing hybrid photosensitive modules is presented. The photosensitive modules in the form of hybrid crystal assemblies of PFC matrices and multiplexers are located on a substrate with minimal gaps between the crystals of not less than 10-20 microns. The block technological route for the manual assembly of a standard

(SD) dimension of 4 modules of 384×288 format, placed on a common switching substrate in a single housing. Experimental block method of photosensitive modules assembly in MFP includes the following main operations: preparation of module surfaces and switching substrate for bonding, application of vacuum adhesive, positioning of photosensitive modules with clamping from above using micromanipulators and control of gaps between crystals using a microscope.

However, specific development of technological approach to creation of the specified FPU for cases of high and ultrahigh dimensionality, when photodetector includes more than 4 modules, is absent here; the gap between modules is not less than 10-20 microns, which is unacceptable in case of small dimensions of PFC, thus image losses in "blind zone" become significant: data volume for subsequent processing is reduced, reliability of video signals processing is reduced and quality of formed images is degraded. Besides, in the technology of manufacturing of this FPU, module clamping is performed from above, using micromanipulators, which is unacceptable in case of IMBP and other monolithic FPUs; even for hybrid microcircuits clamping from above is often unacceptable.

1.2. THE MOST PROMISING APPROACHES TO THE CREATION OF MFP

In [49], a research model of a 14336×10240 MFP, performed on the basis of a matrix of 7×5 submodules with each 2048×2048 elements (Figure 8), is considered. This MFP is one of the possible technological approaches to manufacture MFPs of high and ultrahigh dimensionality. The submodule is a hybrid assembly of a PFC matrix crystal and a multiplexer crystal connected using indium microtubes. As part of the MFP, submodules in open packages are aligned in all coordinates using mechanical elements and fixed in a given position on the carrier plate.

However, this MFP (see Fig. 8) has significant drawbacks: MFP consists of submodules of the same spectral range, the alignment of submodules is performed using mechanical fastening elements, which together with the presence of open housings determines significant "blind spots" of this technological variant of MFP.

In [50] an experimental method of hybrid FP fabrication is presented, which is one of the most promising technological approaches to fabrication of MFPs of high and ultrahigh dimensionality. The indicated variant of hybrid FP creation consists in the fact that in the process of manufacturing, reference marks are created on the multiplexer crystals, retainers are made, external reference marks carriers are made, external reference marks are placed on external reference marks carriers, epoxy glue is injected in the specified places, Orientation of the upper crystal relative to the lower one in the "XY" plane is performed with the help of micromanipulator and microscope by shifting the crystal of the PFC matrices to coincidence of the external reference marks with the reference marks on the multiplexer crystal, the vertical load on the combined multiplexer crystals and PFC matrices is increased up to a certain value.

The methodological basis of hybrid FP fabrication is partially applicable in the development and research of the basic technology of fabrication of MFPs of high and ultrahigh dimensionality. However, the indicated method is directly focused only on the microassembly of two crystals: the PFC matrix and the multiplexer, which limits the format of the fabricated

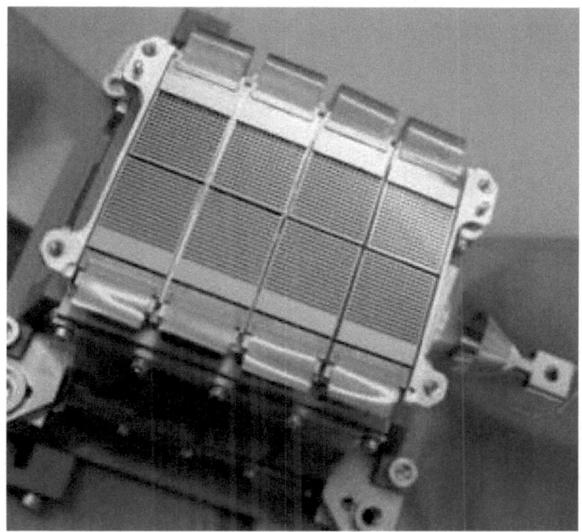

Figure 7 - MFP as a 4×2 submodule matrix [48].

Figure 8 - MFP as a matrix of 5×7 submodules [49].

FPs; direct application of this method for creation of MFPs of high and ultrahigh dimensionality is hampered by unworked issues of formation of extremely smooth faces of submodule crystals with minimal damage areas of semiconductor materials and multilayer nano- and microstructures, precision positioning and alignment of crystals during microassembly of submodules in MFPs. In practice, pressing from above is a damaging effect even for hybrid submodules, and for monolithic submodules, especially for microbolometer receivers, such technological operation is not acceptable at all due to the destructive effect on sensitive elements (SEs).

Semi-automatic method of creating experimental models of MFP in the form of 2 × 2 submodule matrix is as follows: four crystals of PFC matrices are successively transferred to the working surface of the setup using a semi-automatic micromanipulator, multiplexer crystals are successively overlaid on the PFC matrix crystals, multiplexer crystals and PFC matrix crystals are connected using indium micropillars, a common substrate is placed over the multiplexers, which ensures the strength of the hybrid assembly as a whole (Figure 9) [51]. However, the above technological approach creates large blind spots and is applicable to MFPs consisting of 2×2 submodule, 1×2 submodule, or 2×1 submodule matrices; the location of contact pads on two sides of submodule crystals also limits the dimensionality of the MFP.

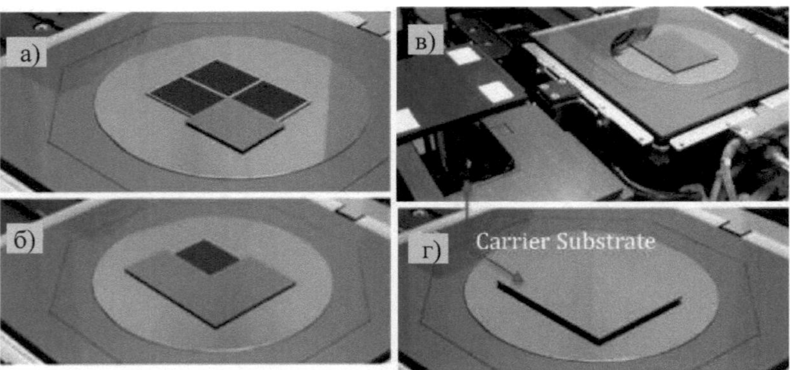

Figure 9 - Semi-automatic method of MFP creation [51].

1.3. ADDITIONAL RESULTS OF THE SCIENTIFIC REVIEW

In [55, 56] mosaic FPs based on single PFCs are considered. In [57] a multichannel (2-3) FPU with readout circuits based on operational amplifiers is presented; the distance between the axes of the element lines is 4 mm, which differs significantly from the level of technological solutions proposed in [1, 9, 10]. In [58] a single optical lens is used for all submodules, and the readout of photo signals in MFP is performed by non-switched summation of signals of a 4×4 element matrix. In [59], "jumpers" between submodules were applied. In [60], a massive multilayer device is proposed; plates with a mosaic of holes are used as layers . The approach is far from the technological solutions proposed in [1, 9, 10]. In [61], a composite array of lenses is used along with the MFP. The design and technology are far from the technological approaches proposed in [1, 9, 10].

The technological solution proposed in [62] is applicable only to linear photodetectors. The materials given in [63-65] are methodologically and by achievable results far from the technological approaches proposed in [1, 9, 10].

An additional method of forming submodule crystal faces is applicable to MFPs [28]. However, a separate technological operation by itself does not provide the creation of MFPs of ultra-high dimensionality with maximum image conversion efficiency [1, 9, 10]. The fabrication method of thin-film multilevel boards for multichip devices, hybrid integrated circuits and crystal microassemblies in combination with the technological approaches proposed in [1, 9, 10] can have a positive "cumulative" effect [66].

Information materials of inventions in patents and applications for
The patents reviewed in [55-96] are methodologically and by the achieved results far from the technical solutions and technological approaches proposed by the authors in [1, 9, 10].

2. SUMMARIZED RESULTS OF THE LITERATURE REVIEW

As a result of the analytical review, the most promising fundamental approaches of the previous level of mosaic technology, presented in [30-37, 48-54], are identified.

2.1. METHODOLOGY FOR ANALYZING THE EFFICIENCY OF IMAGE TRANSFORMATION IN MFP

The purpose of this section is to analyze and further develop mosaic technology for creating high-dimensional FPs and integrated circuits for reading photo-signals of submodules, ensuring minimization of the "blind zone" size between the edge PFCs of adjacent crystals and increasing the efficiency of image conversion in MFPs.

In the most general form for quantitative estimation of image losses it is possible to enter efficiency (ηeff) of image transformation in MFP: ηeff $= 1 - \varepsilon mfpa$, where εmfpa - inefficiency of image transformation in MFP. Inefficiency (εmfpa) of image conversion can be represented as [1, 9, 10]:

$$\varepsilon mfpa = N1\ \varepsilon1 + N2\ \varepsilon2 + N3\ \varepsilon3 + N4\ \varepsilon4,$$

where $N1$ is the number of submodule crystals butted on one side (the first type of submodules); $N2$ is the number of submodule crystals butted on both sides (the second type); $N3$ is the number of submodule crystals butted on three sides (the third type); $N4$ is the number of submodule crystals butted on four sides (the fourth type); εj - relative losses of input image elements in submodule *of j-th* type, reduced to full number of input image elements on the joint photoreceiving matrix of MFP (relative inefficiency of image transformation in submodule *of j-th* type in MFP); $\varepsilon j = {}_{NLj/NRma}$, $j=1, 2, 3, 4;$ ${}_{NLj}$ - amount of input image elements lost in submodules of the first ($j=1$), second ($j=2$), third ($j=3$) or fourth ($j=4$) types, ${}_{NRma}$ - total amount of input image elements, focused on the joint photoreception matrix of MFP. There are certain peculiarities of using submodules of different types [9, 10].

In simpler cases, when, for example, MFP consists of unified submodules with the same element pitch, image loss in MFP can be characterized by the number of elements (${}_{NL}$) of the input image projected on the "blind zones" of MFP, or the image loss coefficient equal to 1/NL [1, 9, 10].

The number of elements in the MFP "blind zone" can be expressed as follows [1, 9, 10]:

$$_{NL} = kx\,M\,(N/n - 1) + ky\,(n + kx)\,(M/m - 1)\,(N/n - 1) + ky\,n\,(M/m - 1),$$

where n is the number of PFCs by "X" coordinate of submodule photoreception matrix, m is the number of elements by "Y" coordinate of submodule matrix, Lpx is the step of PFCs by "X" coordinate of submodule matrix, Lpy is the step of elements by "Y" coordinate of submodule matrix, N *is* the number of PFCs by "X" coordinate of united MFP matrix, M - number of elements by "Y" coordinate of MFP matrix, Lbx - distance by "X" coordinate between edge PFEs of adjacent submocules, Lby - distance by "Y" coordinate between edge PFEs of adjacent submodules, $kx = Lbx/Lpx$ - number of elements by "X" coordinate in "blind zone" of MFP, $ky = Lby/Lpy$ - number of elements by "Y" coordinate in "blind zone".

The number of elements in the 'blind zones" increases linearly with increasing the format of the photodetector matrix and decreases with increasing the size of the PFE. At high and ultra-high format of MFP photodetector matrix and large PFE step the image conversion efficiency in MFP

can remain high, but the number of FFEs in the ' blind zones" in this case can reach critical values (Figure 10) [1, 9, 10].

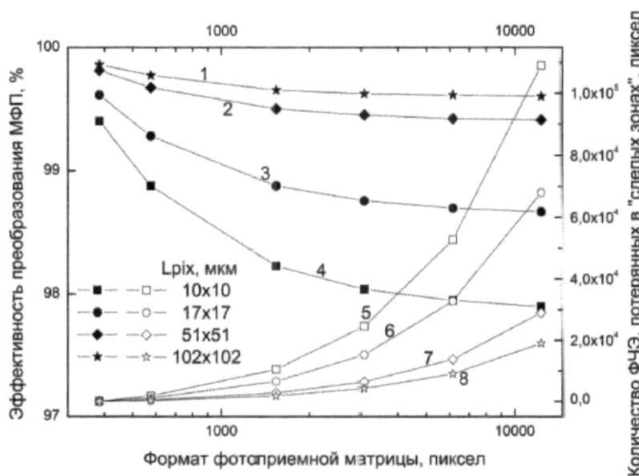

Figure 10 - Dependences of MFP efficiency (1-4) and the number of PFCs (5-8) lost in "blind spots" on the dimensionality of the photoreceiving matrix.

A number of circuits for submodule addressing block (BA) that minimize the MFP "blind spot" were synthesized and proposed [1, 36]. The system rationality of the BA scheme based on two shift registers in comparison with the decoder circuit is analyzed. At construction BA on the basis of the decoder circuit the number of transistors and control buses grows as the format of the photodetector matrix of submodule increases and at achievement of the certain format the advantage for the scheme on the basis of two registers is provided [1, 36].

The size of the "blind zone" perpendicular to the docking lines is determined by the size of the damage area of the semiconductor material arising in the process of separation of wafers into crystals, technological gaps between the docked crystals and technological area of placement on the crystals of schemes for reading photonsignals. The size of the "blind zone" of the MFP along the docking lines of adjacent submodules increases with increasing the format of the combined photodetector matrix and the PFE step.

When creating 1920×1080 MFPs based on a 3×2 matrix of 640×540 submodules with PE spacing from 17 to 51 μm, the number of elements in the "blind spot" will not exceed one or two in each row and column of the photodetector matrix of submodule crystals [1, 9, 10].

2.2. MFP OPERATING PRINCIPLE

The image of the imaged scene is projected onto the focal surface3 of the MFP photoreceiving matrices. The arrays of PFC submodules can be of different shapes; the number of elements is determined by the required spatial resolution of the MFP. It should be especially noted that in many cases optical systems based on lenses are not used in mosaic FPUs; in some cases, mirror systems with specified parameters, as well as optical capacitors or transducers, are used.

In MFPs, submodules of given ranges can operate in parallel, providing maximum readout frequency of ultra-high-dimensional image frames at standard frequency of submodules, which determines relevance of mosaic technology as applied to cooled IR FPs based on MSCN photodetectors and CRT photodiodes and uncooled microbolometer radiation receivers of IR and THz ranges [1-10, 27, 28, 30-37].

2.3. MODERN FUNDAMENTAL GROUNDWORK FOR CREATING MFP WITH MAXIMUM CONVERSION EFFICIENCY

PICTURES

The purpose of this section is to depict the development and optimization of technological operations of precision laser scribing of instrument wafers, separation of wafers into crystals and microdocking of MFP submodules.

In 2013, a basic technological block of laser scribing operations was developed and investigated as part of a prototype of a precision technology of silicon crystal microassembly in MFP [30-36]. Width

The "blind zone" between the edge PFCs of neighboring crystals within an IGBP along the vertical line of the crystal junction is determined by the size of both the technological part of the "blind zone" and the topological area of placement on the chip of the reversible BA submodule of the IGBP along the corresponding coordinate, and along the horizontal line of junction depends only on the possibilities of the precision separation of the plates into crystals. The technological part of the "blind spot" is determined by the width of the damage area of the semiconductor material, arising in the process of separating wafers into working crystals, the roughness of the edges of the docked crystals and the gap between them.

As part of the mosaic technology of MFP creation, an experimental technological operation for microassembly of MIBP crystals was applied. The plate was placed under a microscope on a table with a vacuum line. An LMPP crystal with a minimum thickness of glue layer applied to the non-working side was placed in a given place with a technological hole, and then held until the glue polymerization with the help of vacuum. After that, the next MIBP crystal was installed similarly [30-36]. However, when using the above technological approach to create the MFP, the total size of the technological part of the "blind zone" between the edge FFEs of adjacent submodule crystals was about 30 μm (Figure 11) [30-36].

To date, numerous studies based on a comprehensive literature review and experiments have optimized the mosaic technology (Figure 12), including the following upgraded operations: laser scribing, separation of semiconductor wafers into individual crystals precision laser formation of butted sides of submodule crystals with a total blind spot size between adjacent submodule marginal   PFEs   of no more than 13 μm [1, 9, 10, 37-46]. Particularly carefully designed: separation into wafer crystals and double-sided laser formation of docked sides of submodule crystals for high MFP dimensionality (Figure 13) [10, 37-44]. Precision laser shaping of the docked sides of submodule crystals provides a total blind area between the edge FFEs of adjacent submodules of no more than 13 μm (Figure 14) [1, 9, 10, 37-44, 46].

As a result of the study of the methodology of formation of submodule crystal edges to create MFPs of high dimensions, a laser method of double-sided groove formation was developed, with optimized parameters of energy density distribution on the surface of the semiconductor material, the method of edge chipping of the crystal was selected [1, 9, 10, 37-44].

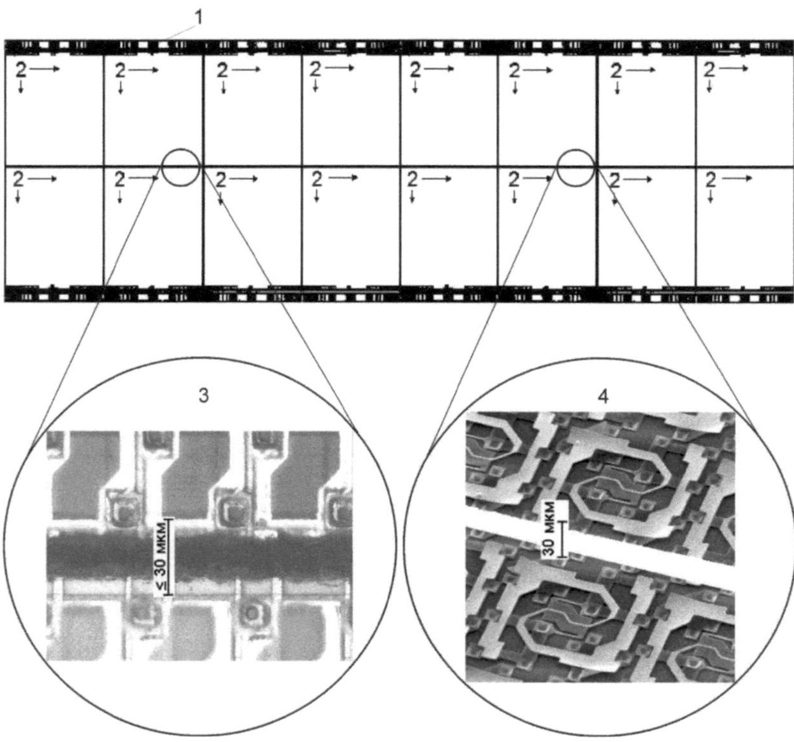

Figure 11 - Principle of construction of a microbolometer MFP (*1*) with format up to 3072×576 and more for IR and THz ranges: *2* - submodules, *3* - docked crystals of IR submodules with specified size of technological "blind spot", *4* - technological version of THz submodule; arrows show single principle of scanning of the MFP as a whole.

Figure 12 - Optimized mosaic technology, creating MFPs as an array of 3 × 2 submodules.

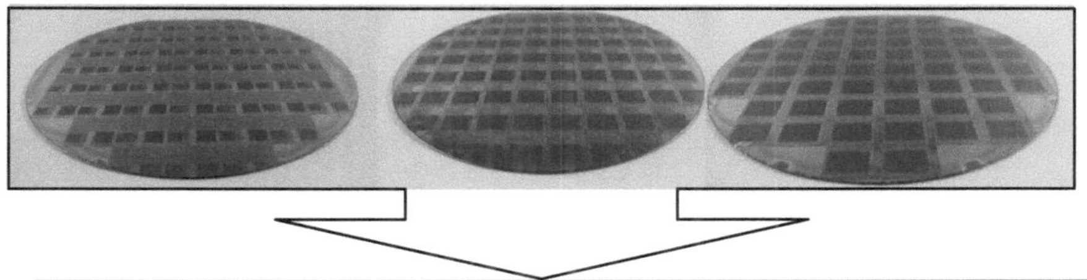

laser scribing of wafers, separation into crystals, precision laser formation of butted sides of submodule crystals

Microassembly of submodule crystals in MFPs with a gap of ≤ 2 μm

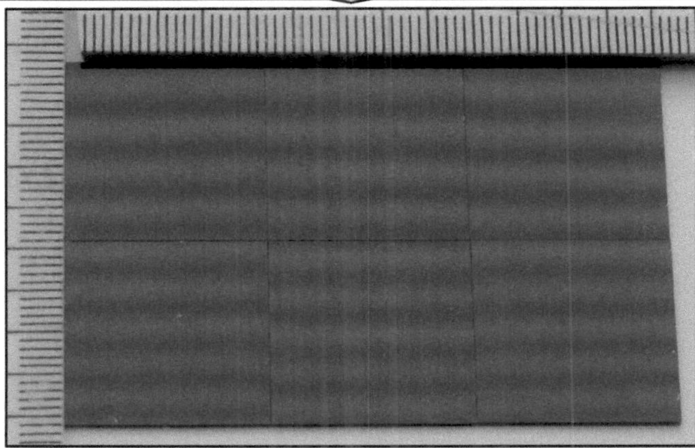

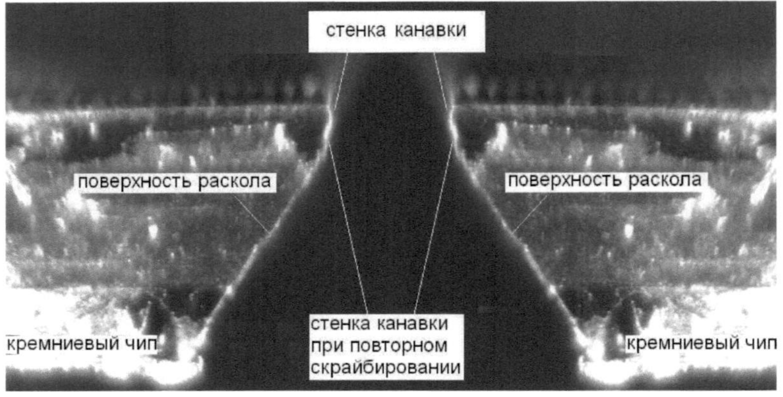

Figure 13 - Photographs of precision-formed for subsequent microdocking surfaces of submodule crystals, side view.

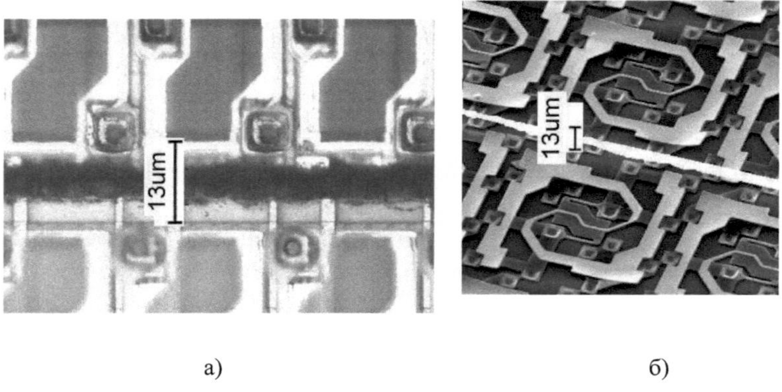

а)   б)

Figure 14 - Photographs of IR (a) and THz (b) MIBP with the indicated total blind zone width between the edge FFEs of adjacent crystals.

Laser formation of the crystal edge is performed by radiation with a given wavelength in multipass mode (120 µm/s), with a given angle of inclination of the optical axis of radiation to the surface normal [28, 37, 41]. In this method of formation, there is no material melt on the crystal surface [28, 37, 41]. The upgraded method of crystal edge formation with the selected method of edge chipping provides an even (roughness of about 1 µm) perpendicular to the planar side of the submodule crystal and allows adjacent submodules to be aligned with minimal gaps [1, 9, 10, 37-44].

The breakthrough scientific and technological result in the technology of MFP creation is based on a single complex of technological operations blocks: placement of frameless submodule crystals butt to butt directly on the carrier plate with minimal gaps between crystals; precision formation of crystal faces with provision of minimal damage areas of semiconductor materials and multilayer nano- and microstructures on the butt sides of crystals; application of vacuum gripper, support bar in the form of a rectangular frame and semi-automatic or automatic micromanipulators for positioning submodules in the MFP manufacturing process.

3. OPTICAL METHODS FOR MAXIMIZING THE EFFICIENCY OF IMAGE CONVERSION IN ULTRA-HIGH DIMENSIONAL MFP TECHNOLOGIES

The purpose of this section is to develop optical methods for achieving minimum dimensions of the "blind zones" between the edge FFEs of adjacent submodule crystals in the technology of MFP creation and, accordingly, to provide the ultimate efficiency of image conversion in MFPs of extra-high dimensionality.

3.1. OPTICAL METHODOLOGY FOR ACHIEVING THE MARGINAL EFFICIENCY OF IMAGE CONVERSION IN MFP EXTRA-HIGH DIMENSIONAL

Let us consider some variants of technological design of MFP of ultrahigh dimension based on a single carrier plate with the use of frameless (hybrid and monolithic) submodules [1, 9, 10, 36-46]. The basic MFP design with maximum image conversion efficiency consists of a matrix of n×m frameless photodetector submodules, where n = 1, 2, 3, 4, 5, 6, 7, 8, ... ∞, m =1, 2, 3, 4, 5, 6, 7, 8, ... ∞, (n ≠ m if n = 1 or if m = 1), set junctionbutt to butt on a single carrier plate [1, 9, 10, 36-46]. Submodule crystals are designed with minimal (5 - 8 µm) damage regions of semiconductor materials and multilayer nano- and microstructures on the docked sides of the submodule crystals. Frameless submodules are placed with the photodetector side either up or down on the carrier plate with minimal (no more than 2-3 µm) gaps between the crystals. In the latter case, the carrier plate is made of material transparent in a given spectral range. The mounting of frameless submodules on the carrier plate is performed by a certain amount of vacuum heat-conductive retaining material; if necessary, retaining material transparent in a given spectral range is selected.The second version of MFP differs by the fact that the edge PFCs of the docked sides of submodules are smaller in area relative to the size of PFCs inside the matrix, and the corresponding correction of PFC signals is provided at the subsequent processing. In MFPs of extra-high dimensionality according to this variant the given step of PFCs in the area of jointing crystals of submodules is kept and losses of the visualized information are excluded.At application of the basic variant of MFP design together with the second variant it can be insufficient to provide the subsequent correction of PFC signals on system considerations. In such a case, the alignment of PFC signals in

MFP is provided by increasing the photon flux acquisition area, coming to the reduced edge PFCs of docked sides of submodule crystals, up to the value of PFC acquisition area inside the submodule photoreceiving matrix. In this connection, additional technological options for creating ultra-high-dimensional MFPs have been investigated, including the design of microdocking areas of adjacent submodule crystals using optical methodology to achieve maximum image conversion efficiency   in MFPs [1, 9, 10, 38-44].

The third version of the MFP   differs in that over each
The individual integral bifocal lens of the corresponding spectral range is placed on each reduced edge PFC of the submodule, focusing radiation from the area equivalent to the area of the PFC inside the photodetector matrix of the submodule (Figure 15) [1, 9, 10, 38-44].

The fourth version of MFP differs in that over the "blind zones" optical prisms of the corresponding spectral ranges and linear sizes are placed, thus eliminating the "blind zone"; the cross sections of the used optical prisms can be of different shapes, as shown in Figure 16 (a, b, c, d, e) [1, 9, 10, 38-44].

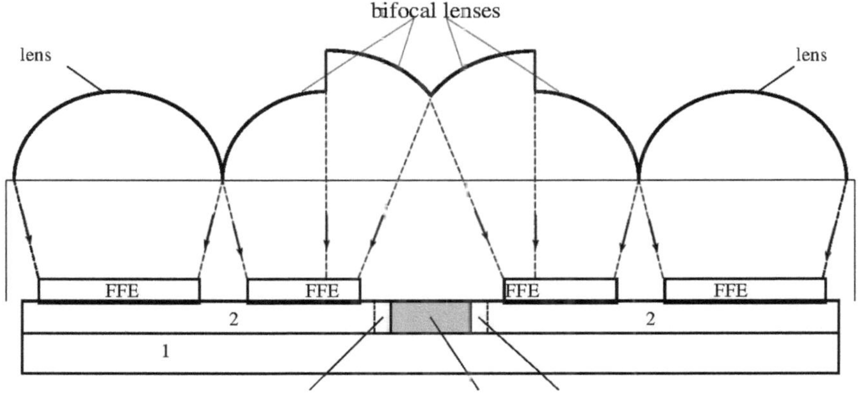

area of material damage 7 area of material damage

Figure 15 a - Technological areas of microdocking of adjacent MFP submodules with individual integral, bifocal lenses of corresponding spectral ranges (basic principle, sketch) [10].

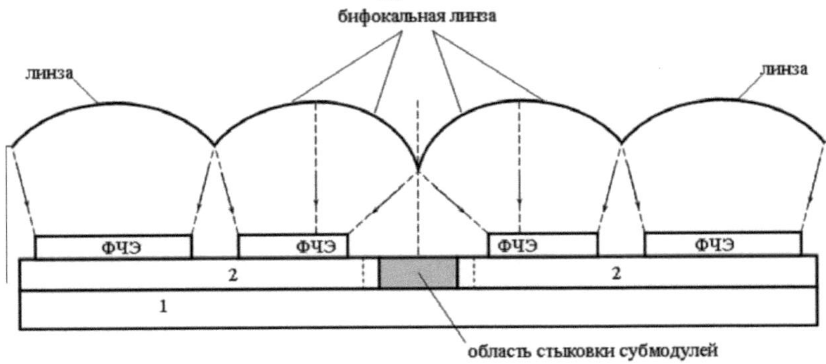

Figure 15 b - Technological areas of microdocking of adjacent MFP submodules with individual integral, bifocal lenses of corresponding spectral ranges (variant of realization, sketch) [10].

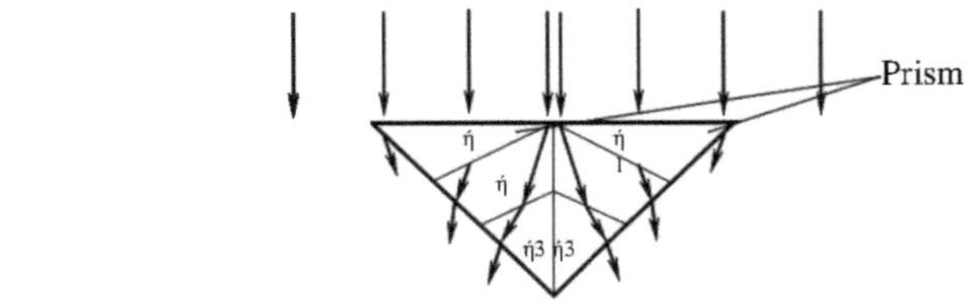

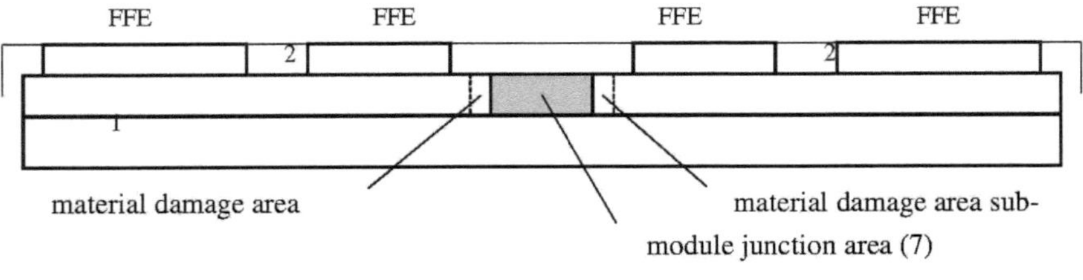

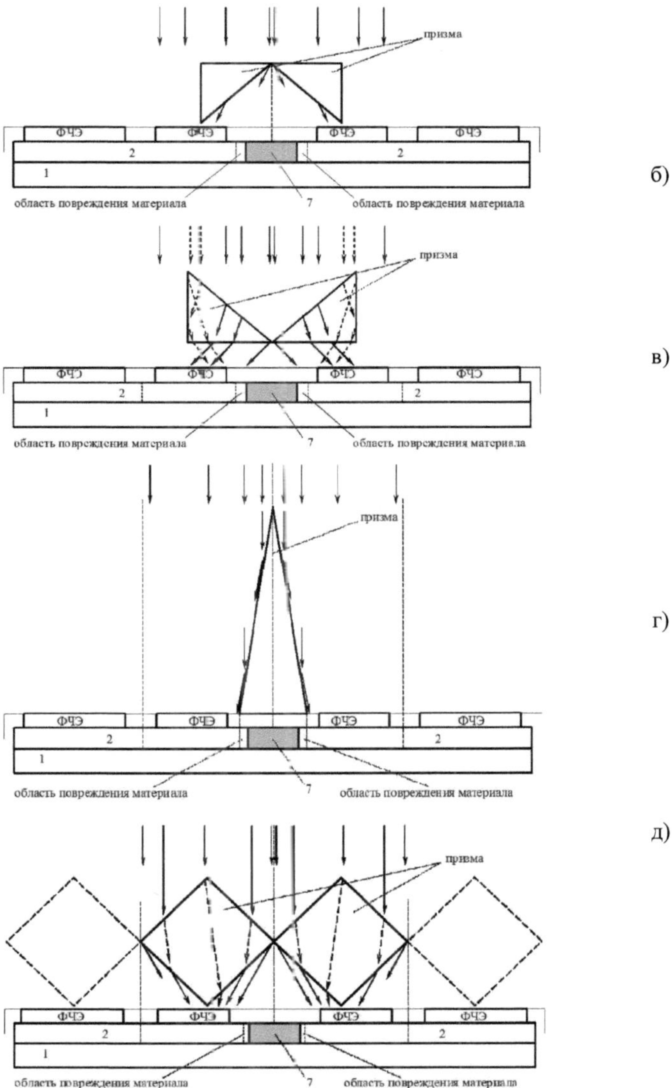

Figure 16 - Technological areas of microdocking of adjacent MFP submodules with optical prisms of the corresponding ranges (sketches) [10].

3.2. GENERALIZED CONCLUSIONS ON OPTICAL METHODS FOR MAXIMUM EFFICIENCY IMAGE TRANSFORMATIONS IN MFP

Technical results of the proposed variants of MFP technological design are, firstly, achievement of maximum efficiency of images transformation in MFP by limiting the number of elements in "blind zones" between edge PFC of adjacent submodules, and in some cases - complete elimination of "blind zones" with preservation of a given PFC step in the area of joint of submodule crystals; secondly, increasing the detection probability of visualized point targets and improving the quality of images of extended objects; Third, an increase in the collection area of photon flux coming to the reduced edge PFCs of submodules up to the value of PFC collection area inside submodule matrices; fourth, providing MFP microminiaturization by placing frameless submodules on a single carrier plate with minimal gaps between adjacent crystals and corresponding expansion of the MFP industrial and scientific application area. Technological variant of MFP, where "blind zones" are optically overlapped by edge FFE of adjacent submodules, eliminates video information losses in each frame and provides achievement of the maximum (100%) image conversion efficiency.

4. TECHNOLOGICAL PROTOTYPING OF MFP
WITH MAXIMUM IMAGE CONVERSION EFFICIENCY

The purpose of this section is to research and develop technological approaches to the design of IMBP submodules, which in a given way minimize the size of the BA on the corresponding coordinate when placing the contact pads on one side of the submodule chip.

Achievement of maximum efficiency of IR and THz images conversion in MFP from this point of view lies in development of basic variant of submodule technological design (Figure 17). Based on the analysis of the principles of creating circuits for reading microbolometer (MB) photosignals, a technological prototype of the submodule was synthesized, which minimizes the size of BA on the corresponding coordinate when placing the contact pads on one side of the crystal in a specified manner.

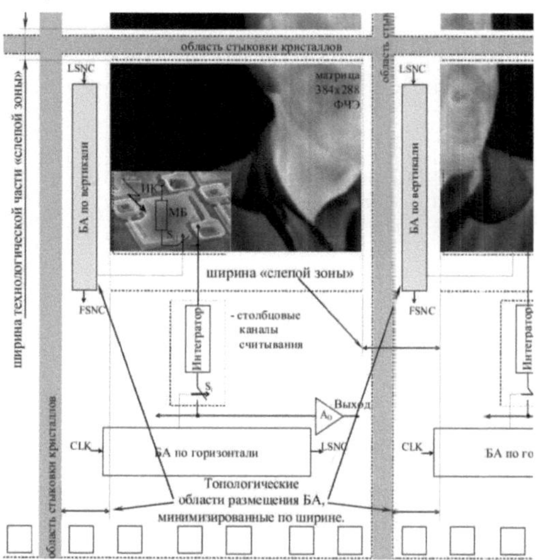

Figure 17 - Basic structure of the MFP with submodules in the form of IMBPs, the variant of IR image arrangement on the PFC matrices of submodules and indicating the locations of the "blind zone" and its technological part; *Vhi* - bias voltage of MB, $s_{i,j}$ - addressing key in the cell, s_i - column addressing key, Ao - output node.

5. NOISE EQUIVALENT TEMPERATURE DIFFERENCE FOR CONGRUENCE OF ULTRA-HIGH DIMENSIONAL MFP

5.1. METHODOLOGY FOR NETD IC MFP ANALYSIS

As presented above: mosaic technology is a promising fundamental basis for achieving ultra-high dimensional IR FPs (see Figure 12) [1, 9, 10, 38-44, 46]. The application of a noise equivalent temperature difference (NETD) for the congruence of different IR MFPs combines the simplicity of the calculation and the clarity of the analysis of the expected signal-to-noise ratio at a given input temperature difference of 1 K [2-9, 38-46].

The temperature resolution of IR MFPs in the spectral range of 8-14 μm was analyzed as a function of various technological parameters of multiplexers and PFCs, namely as a function of the *ROA* parameter, quantum efficiency η, PFC size *lpix*, charge capacity of the QROIC integrator in each multiplexer channel, and wavelength λmax of the maximum of the PFC spectral sensitivity characteristic. Typically, multiplexers are created in the form of an array of photocell processing cells and a system for reading the transformed cell signals to a common output [2-9].

NETD values were estimated as the ratio of the noise of the system including the IR PFC and multiplexer readout channel to the temperature sensitivity of the IR PFC at a given quantum efficiency η:

$$NETD = \frac{U_n}{\dfrac{\partial U_{sig}}{\partial Q}\dfrac{\partial Q}{\partial T}} \approx \frac{U_n}{\Delta}\frac{\Delta_{}}{{}_{sig}} \quad , \qquad (1)$$

where U_n is the noise voltage of the system including the IR SPE and the readout channel multiplexer; $\partial U_{sig}/\partial Q$ is the volt sensitivity of the PFC; Q is the input photon flux; $\partial Q/\partial T$ is the temperature gradient; ΔT is the temperature difference at input of the optical system and ΔU_{sig} - signal voltage difference for a given difference ΔT [2-9, 12, 17, 39, 43, 45, 46, 97].

The noise voltage (U_n) of the IR PFC-multiplexer readout channel system, reduced to the integrator capacitance, was calculated as a function of the parameter ROA or the dark current I_{dark} at the background temperature T_{bkg} = 300 K and aperture angle =30o. A generalized expression illustrating the calculation of U_n can be written as follows

$$U_n = \sqrt{\left(\frac{4kT}{R_{det}} + 2qgI_{inp}\right)\frac{\eta_{inj}^2 t_{int}}{2C_{int}} + \frac{-}{3C_{int}^2} + \int_{\Delta f}\frac{t_{int}}{R^2_{det}C_{int}}\left(\frac{8kT}{3g_m} + \frac{A_m}{f^{\alpha}}\right)df + \frac{2kT}{C_{int}} + \int_{\Delta fa}\left(\frac{8kT}{3g_{ma}} + \frac{A_a}{f^{\alpha}}\right)df}$$

,

$$(2)$$

where R_{det} - differential resistance of PFC; I_{inp} - total current of PFC at operating point; η_{inj} - input photoinjection coefficient; C_{inp} - input node capacitance; C_{int} - integrator capacitance; t_{int} - accumulation time; g - photoelectric coefficient of MCN [5, 6, 98]; $g=1$ - for CRT photodiodes; g_m - steepness of input transistor g_{ma} - steepness of the active amplifier transistor; A_m - noise spectral density of the input transistor at 1 Hz; A_a - noise spectral density of the amplifier transistor at 1 Hz; Δf - integrator bandwidth; Δf_a - amplifier bandwidth [2-9, 12, 17, 39, 43, 45, 46].

When creating IR FPs, it is usually required to provide a "background limiting" ("BLC") mode, when the FP parameters are determined by the noise of the background radiation. In this case, using formula (2) it is not difficult to obtain the following expression describing the dependence of the temperature resolution of IR FP on the parameters of the chosen multiplexer technology:

$$NETD\sqrt{\frac{TOX}{EP}} ,$$

$$(3)$$

where TOX is the thickness of the working dielectric, EP is the supply voltage of the technology, selected for multiplexer fabrication [2-9, 39, 43, 45, 46].

In some cases it is difficult to provide the "OF" mode, for example, because of the high energy consumption for cooling the photodetector to the required operating temperature. In such cases, although the input coefficient of the signal photocurrent remains high enough, the differential resistance of MSCN photodetectors significantly decreases and the temperature resolution is determined by 1/f - noises of the input transistors in the multiplexer cells. At the same time, the analytical dependence of the estimation of the equivalent temperature difference noise on the technological parameters changes slightly:

$$
NETD \sim \begin{cases} \dfrac{TOX}{\lambda_{-M}}\sqrt{\dfrac{\lg(\dfrac{Tobs}{Tint})}{}}, & \text{for } \Delta N \text{ Dependent } 1/f \text{ models - noise,} \\[2em] \dfrac{\sqrt{OS}}{\lambda}\; TOX \; Tint \sqrt{\lg(\dfrac{Tobs}{Tint})}, & \text{for } \Delta\mu \text{ Dependent } 1/f \text{ models - noise,} \\[1em] c \quad MOS \; EP \end{cases}
\tag{4}
$$

where λCMOS is the characteristic topological size of the silicon technology, t_{obs} is the observation time of the photon signal related to the minimum frequency of 1/f-noise as follows: $f0=1/t_{obs}$; $tint$ is the accumulation time of the photon signal; ΔN is the fluctuation of the carrier number due to the capture and release of carriers by traps; $\Delta\mu$ is the fluctuation of carrier mobility due to their scattering on phonons [2-9, 39, 43, 45, 46].

Qualitative estimates of NETD considering *(2)* and *(3)* are shown in Figure 18, curve *1* corresponds to $R0A$ = 100 Ohm cm2, curve *2* - $R0A$ = 200 Ohm cm2, and curve *3* - $R0A$ = 2000 Ohm cm2. Thus in case of transition from technology 0,6 μm to 0,35 μm the combination of accompanying changes of parameters of the readout circuit, set by the chosen technology, such as thickness of working dielectric and supply voltage, leads (when maximum capacitance and accumulation time are used in the circuit) to decrease in its volt sensitivity in 1.5 times, while the noise voltage of this system decreases only in 1.3 times, which creates a small local maximum on the NETD dependence of the multiplexer technology design norms [5-9, 39, 43, 45, 46]. In case of transition from 0,6 μm technology to 0,25 μm, change of a ratio of parameters of system "FSE on the basis of MSCN - readout multiplexer channel" causes more considerable decrease in a noise voltage in comparison with decrease of its sensitivity and provides transition to area of lower values of the equivalent noise difference (see figure 18) [5-9, 39, 43, 45, 46].

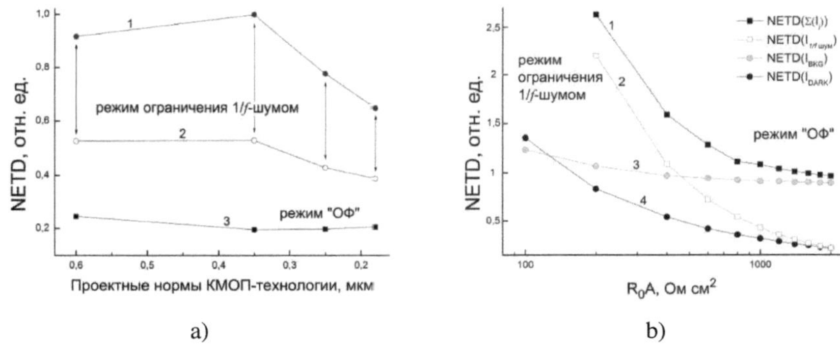

Figure 18 - Dependences of the noise equivalent temperature difference of IR FPs based on MSCN photodetectors with $\lambda_{max=9}$ microns from multiplexer technology design norms (a), from R0A (b): *the curve 1* - NETD taking into account all noise sources, *2* - NETD contribution 1/f - input transistor noise, *3* - NETD share taking into account background radiation, *4* - NETD contribution of dark current.

In the OF mode, NETD is inversely proportional to the square root of the accumulation time *tint* - $NETD \approx 1/(t_{int})^{1/2}$, so in the case where the temperature resolution is determined by *1/f-noise of* the input transistors in multiplexer cells, the ratio of the noise contribution to the NETD value should be checked most carefully, because the direct proportionality: $NETD \approx (t_{int})^{1/2}$ follows from the expression for the $\Delta\mu$-dependent *1/f-noise* model [2-3, 5-9, 39, 43, 45, 46].

To further evaluate the NETD FP in the longwave spectral IR range, a prospective MSCN sensitivity band of 8.5-
9.5 μm [5-9, 39, 43, 45, 46]. The accumulation time was determined from the condition of complete filling of the integrator capacitance (*Cint*) with regard to the supply voltage of the multiplexer technology with frame accumulation (FAC) [2-3, 5-9, 39, 43, 45, 46]. The characteristics of IR FPs have been analyzed for multiplexers with KN with a 640×512 matrix format, four outputs, column amplifier feedback circuit capacitance equal to *Cint*; a cell pitch of 20×20 μm and 0.18 μm technology [4-9, 39, 43, 45, 46].

5.2.   NETD ANALYSIS RESULTS FOR MSCI MFP

The resulting analytical estimates of NETD FPs using expressions *(1)* - *(2)* are illustrated in Figure 19 [2-3, 5-9, 39, 43, 45, 46]. The advantages of using multiplexers with KN for MSCN photodetectors and multiplexers with string accumulation (PN) for CRT photodiodes in the extended 8-16 μm long-wave IR range have been revealed in a series of articles [2-9, 39, 43, 45, 46].

MSCC-based FPs (λmax $\approx$ 9 μm) provide good NETD when the *R0A* parameter is more than 500 Ohm cm2 and the quantum efficiency $\eta \geq 0.1$ [2-3, 5-9, 39, 43, 45, 46]. The NETD parameter of MSCN-based FPs monotonically increases
as the linear dimensions of the PFC decrease [2-3, 5-9, 39, 43, 45, 46]. Charge capacitance of multiplexer integrator: QROIC=CintEP. In typical IR FPs at 10% quantum efficiency the charge capacity of the integrator (QROIC)

is not required more than 14 pCL; at a quantum efficiency of 15% it is sufficient QROIC=17-20 pcL [2-3, 5-9, 39, 43, 45, 46].

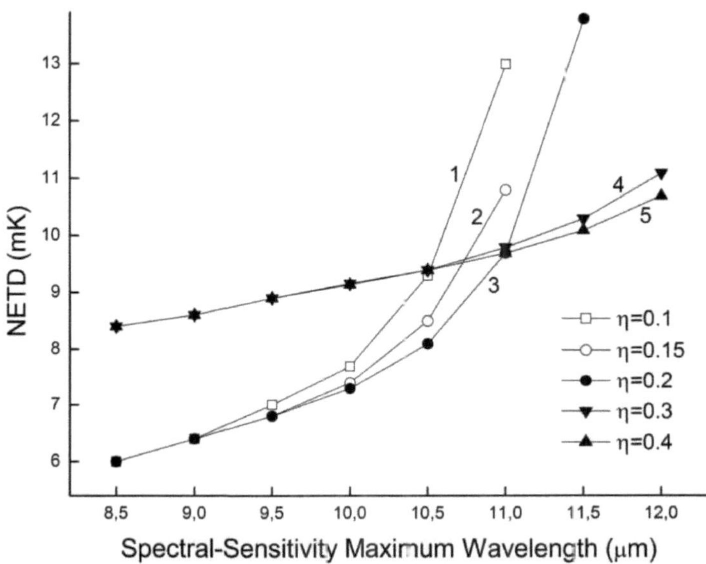

Figure 19 - NETD dependences of IR FPs on the wavelength (λmax) of the maximum spectral photosensitivity of PFCs at different values of η.

The NETD FP values obtained at η = 0.1-0.2 (*curves 1-3*) and at η = 0.3-0.4 (*curves 4, 5*) are illustrated in Figure 19 [2-3, 5-9, 39, 43, 45, 46]. In the wavelength range of the maximum spectral sensitivity $_{max} \approx 8.5$- λ 9.5 μm FP at η = 0.1-0.2 provides a good (6-7) mK temperature resolution, almost independently of the quantum efficiency η, since NETD is determined by background radiation noise [2-3, 5-9, 39, 43, 45, 46]. In the range λmax $\leq$ 10-11 μm, the NETD FP at η = 0.1-0.2 is smaller than the NETD FP at η = 0.3-0.4. Note that the quantum λ efficiency of 0.1-0.2 is more consistent with MSCN photodetectors, while η = 0.3-0.4 is closer to superlattice-based photodetectors. With a further increase in $_{max}$, an increase in NETD is observed with an advantage for larger values of η photodetectors (see Figure 19) [2-3, 5-9, 39, 43, 45, 46].

6. EXPERIMENTAL RESULTS

6.1. Some EXPERIMENTAL DATA ON PROTOTYPE IR AND THC MOSAIC IMBPs

The BA scheme developed in section 4 and patented using 0.5 μm 1P 3M technology provides the minimum topological size of the "blind zone" area of microbolometer MFPs: 20 μm for THz PFCs in 100 μm increments and 26 μm for IR spectral PFCs in 51 μm increments [10, 31-35, 99].

Application of the above experimental results in creating 3072×576 format microbolometer MFPs, for example, based on submodules of 384×288 size provides in the IR spectral range conversion efficiency of images more than 99%, and in the THz range conversion efficiency can reach 100%. Experimental THz images were obtained using basic IMBP structures similar to those shown in Figure 11 and Figure 14 (Figure 20).

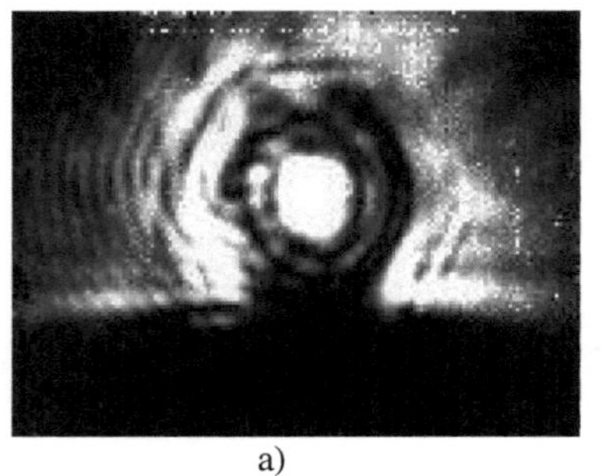 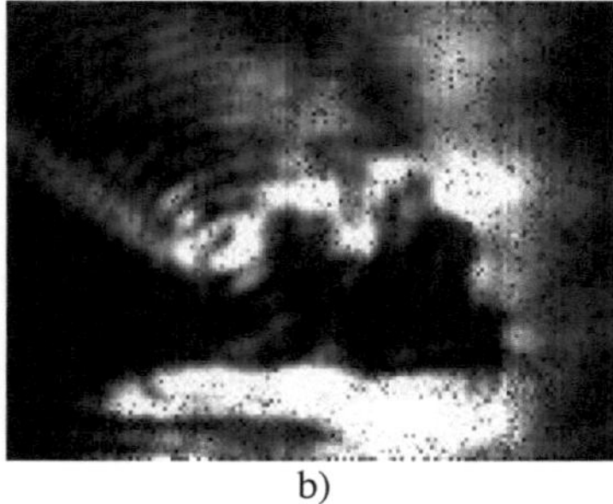

a)   b)

Figure 20 - Images in THz range: metal nut "M8" (a) photographed through 2 mm thick black plastic,
"grooves" of the metal key (b) [31-35].

6.2. RESULTS OBTAINED FOR PROTOTYPE M MFP

Linear (1×288, 1×576) and matrix (128×128, 320×256) technological prototypes of IR MFP submodules are made based on the developed multiplexers and created photodetector crystals [2- 9, 45, 46, 100-105]. Characteristics of some prototypes of multiplexers and IR FPs are summarized in Table 1.

For matrix IR FPs, 128×128 multiplexers ("MHV") with the IP based on CCD/CMOS technology with extremely low dark currents and low noise and 320×256 multiplexers ("MM-1") with the IP of CRT photodiodes and MSCN photodetectors (Figure 21) were created [2-9, 45, 46, 100-102, 105].

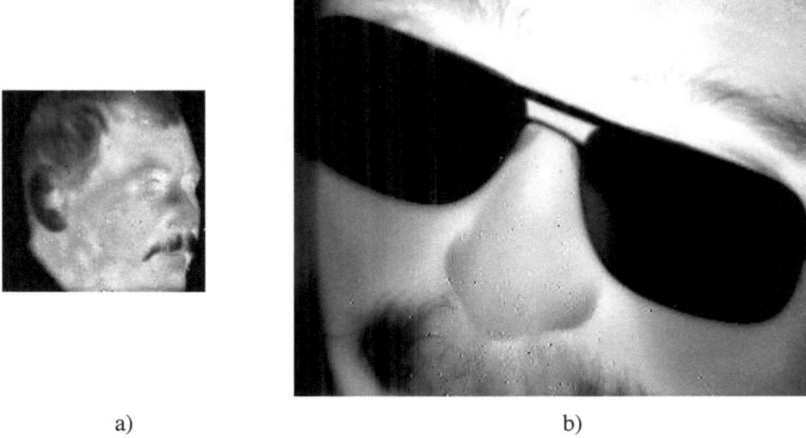

a)   b)

Figure 21 - Examples of 128×128 (a) and 320×256 (b) IR images obtained using MSCN-based FPs [2-3, 5-9, 103, 104].

For matrix IR FPs based on photodiodes two variants of multiplexers with PU were created: 128×128 MMPN-1 with a cell pitch equal to 40×40 μm, and 320×256 MMPN-2 with a cell pitch equal to 35×35 μm (see Table 1). In the extended (8-16 μm) long-wavelength IR range, the PH multiplexers provide better NETD compared to known similar crystals. In contrast to multiplexers with KNs, lower NETD values of FPs with PNs can be achieved in the range of larger $_{max}$ photodiodes [2-4, 7, 8, 45, 100-102, 105].

For linear IR FPs based on photodiodes, multiplexers have been created LM-1, LM-2 and LM-3 (see Table 1) [2-4, 7]. LM-3 multiplexer is compatible with the

4×288 photodetector crystal. These multiplexers with programmable integrator capacitance provide the creation of inexpensive longwave IR FPs for civilian applications, working with increased background radiation and with photodetector crystals with non-standard dark currents [2-4, 7].

The difference of the experimental NETD of the created IR FPs from the theoretical values is determined by the values of sensitivity of the applied PFCs and heterogeneity of the parameters in the field of the photodetector matrices [2-9, 39, 43, 45, 46, 100-105]. Nevertheless, multiplexers provide creation of IR FPs for spectral bands 8-14 µm and 3-5 µm with NETD (14-25 mK) corresponding to the world level [2-9, 39, 43, 45, 46, 100-106].

6.3. EXPERIMENTAL DATA ON MSCI MFP

A variant of MFP with 3840×2160 dimensions can consist of a matrix of 3×3 crystals of submodules, which have the format 1280×720 with the number of MSCN photodetectors equal to 921600 elements; the total number of MSCN photodetectors in MFP - 8294400 elements.

In Figure 22, the achievable NETD values are analyzed for the 7680×4320 MFP dimension with a 10 µm pitch *(curve 1)* or 20 µm pitch *(curve 2)* and a frame rate of 100 Hz, depending on the submodule matrix format and MSCN pitch - photodetectors; conditional reference points are shown in the graph. The required clock frequency *(curve 3)* and the required number of signal outputs *(curve 4)* are analyzed depending on the dimensionality of the submodule matrix. In practice, the achievable NETD of an ultra-high dimensional IR MFP is determined by the thermal resolution of the photodetector submodules, does not depend on the submodule format, the required clock frequency and the required number of signal outputs change. An ultra-high dimensional 7680×4320 MFP based on a combined matrix of 4×4 photodetector submodules of high 1920×1080 format includes 33177600 elements.

Ultra-high-dimensional MFPs, in addition to high performance and efficiency of parallel reading of photographic signals at the standard sub-module clock frequency, provide a virtually unlimited increase in sensitivity by structuring PFCs of large total area, and as a result, a significant increase in target detection range and pattern recognition in the infrared spectral range [1, 9, 10, 39-46]. The MFP format is composed based on the mosaic principle, and the sizes of the structured PFCs are formed by summing the areas of the

PFCs according to the required sensitivity [1, 9, 10, 39-46]. A quantitative analysis of the MFP "blind zones" is performed in subsection 2.1 [1, 9, 10, 39-46].

Mosaic FP technology also provides the formation of multispectral sensitivity characteristic, combining broadband and narrowband parts, using different submodule crystals based on PFC of different types, for example, CRT photodiodes and MSCN photodetectors, sensitive simultaneously in different IR spectral ranges. The CRT photodiodes can provide the broadband portion of the multispectral sensitivity, and the MSCN photodetectors can provide the narrow-band part of the sensitivity characteristic [1, 9, 10, 39-46].

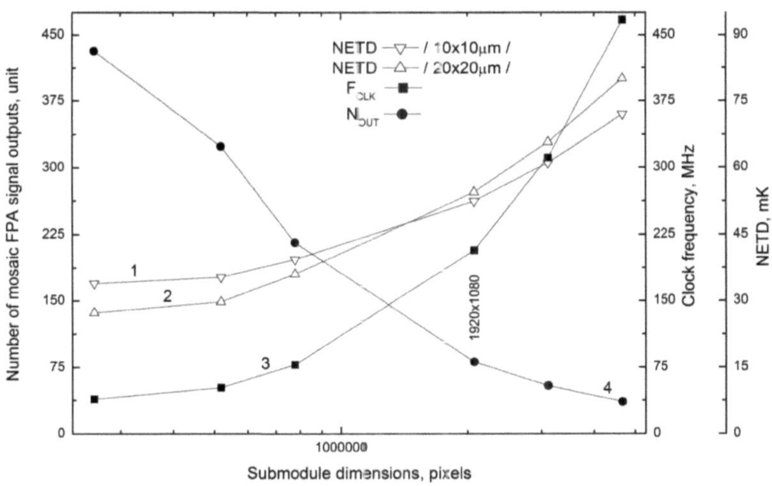

Figure 22 - Dependences of equivalent noise temperature difference, FCLK clock frequency and number of signal outputs (NOUT) of MFP ultrahigh dimensionality on submodule format and readout cell pitch.

Multispectral MFPs can be created on the basis of MSCN submodules of different IR spectral ranges; or on the basis of IR and THz microbolometer submodules used simultaneously. Technological capabilities and ultimate achievable MFP blind spot sizes are given for different materials in Table 2 [1, 9, 10, 27, 28, 37].

Thus, the fundamental technological foundations for the creation of ultra-high-dimensional MHPs with maximum image conversion efficiency have been developed. Basically, the obtained results are applicable to the creation of ultra-high-format MHPs based on microbolometers, CRT photodiodes and MSCN photodetectors, including multispectral high-dimensional FPs.

7. TABLES

Table 1: Characteristics of some prototypes: multiplexers and IR FPs

Format	Step (μm)	Multiplexer	Entrance schematic	Type of FFE	$\lambda_{(\mu m)}^{max}$	NETD (mK)	
						Evaluation	experiment.
320 256 × 128 128	40×40	MM-1	PI	GaAs/AlGaAs QWIPs	9	13-19*	-
	50×50	MXV			8,3	11-14	22**
320 256 × 1 576 × × 1 288	40×40	MM-1	PI	HgCdTe photodiodes	6	6-8	17
					5	7*	25
	35×35	MMPN-2	PI, THE GPU		11	7-9	20
	30	LM-1			10,2	13	14
		LM-2	BPI		8-16	9-19***	-
	28	LM-3	PI				

*Note: PI - direct injection; DPI - subtracting the level of the photo signal individually for each line; BPI - buffered direct injection; *- values at Fframe=120-400 Hz; **-value at 65 K; ***-values atmax=11- 13 µm, Fclock=5 MHz [2, 3, 7, 100, 101, 103, 105].*

Table 2. Fundamental characteristics of the prototype mosaic technology

№	Determining-supporting material	Gap* (μm)	Size* of damaged area (μm)	Size* "blind spot" (μm)	Number of lost elements/step FFE (pcs./μm)
1	Si (multiplexer)	≤ 2	≥ 5	$11 - 13$	1 / 10
2	Si (IMBP)	≤ 2	≥ 5	$1 - 3$**	0 / 10
3	GaAs/AlGaAs lined	≤ 2	≈ 5	$11 - 13$	1 / 10
4	GaAs/AlGaAs with thinning of the substrate	≤ 2	≈ 5	$11 - 13$	1 / 10
5	HgCdTe at GaAs substrate	≤ 3***	≥ 8 [27, 28, 37]	$17 - 19$	2 / 10
6	HgCdTe on Si substrate	≤ 3***	$\geq 5-8$ [27, 28, 37]	$17 - 19$	2 / 10

*Note: *- preliminary experimental results confirming the theoretical values; **- values correspond to the maximum (100%) image conversion efficiency in mosaic IMBPs; ***- reference values for photodetector crystals based on CRT photodiodes [1, 9, 10, 37].*

8. THE ACHIEVED EXPERIMENTAL LEVEL OF MOSAIC TECHNOLOGY

Mosaic principle of creation of IR FPs of ultrahigh dimension based on MSCN photodetectors is shown in Figure 23: "blind zones", areas of damage of materials and gaps between the crystals of submodules are shown. Measurement of the dependences of the p-n junction current on the manufactured laboratory samples from the distance to the groove confirms the minimum size of the damage regions of 5 - 8 μm [1, 9, 10, 27, 28, 37].

The achieved level of mosaic technology for precision submodule microassembly in MFPs is demonstrated in Figure 24: a technological prototype of MFPs in the form of a 3×3 submodule matrix; the gaps between adjacent submodule crystals are no more than 2 μm [1, 9, 10, 27, 28, 37]. The limits of the optimized mosaic technology and the blind spot sizes provided are compared for different defining materials in Table 2 [1, 9, 10, 27, 28, 37].

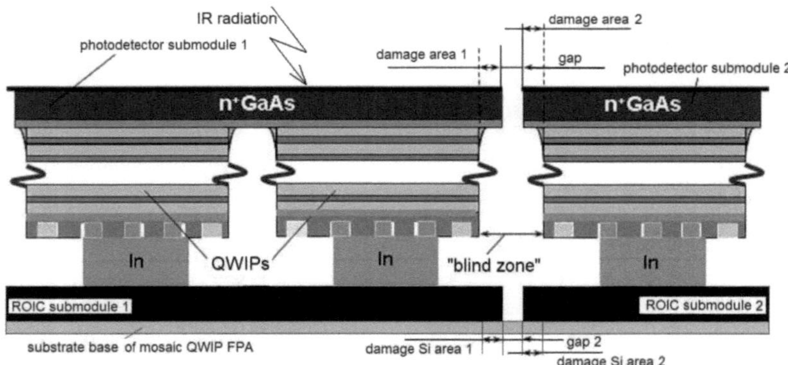

Figure 23 - Technological design of IR MFPs of ultra-high dimension based on MSCN photodetectors, side view.

Figure 24 - Example of a technological prototype MFP in the form of a precision 9-crystal submodule microassembly, the MFP is shown with minimal gaps between adjacent crystals.

9. GENERALIZED FINDINGS

Ultra-high-dimensional MFPs with high image conversion efficiency are used for imaging in different spectral ranges [1, 9, 10, 13, 17, 27-46, 54]. If the gap between the crystals of adjacent submodules is 10-20 μm, then for the case where silicon crystals are the defining ones, the minimum size of the technological "blind spot" of MFP would be 20-30 μm, which corresponds to a loss of 2-3 FFE at 10 μm steps in each row and column [1, 9, 10, 25, 28, 30- 44].

The minimum size of the technological "blind area" of MFPs of 20-30 μm is only possible if precision formation of docked crystal sides with the minimum damage area of semiconductor materials and multilayer nano- and microstructures is applied [1, 9, 10, 28, 37-41, 43, 44].

In the case where other materials, such as HgCdTe, are the determining ones, the minimum distance between the edge CRT photodiodes of adjacent submodules would be 26-36 μm, which corresponds to a loss in each row and column of 3-4 PFCs with the same pitch and the same condition of applying precision formation of docked crystal sides [1, 9, 10, 27, 28, 37].

When using double-sided laser shaping of the docked sides of crystals to ensure minimal damage area, the gap between silicon crystals of adjacent submodules or matrix crystals of MSCN photodetectors, such as those based on GaAs/AlGaAs, is no more than 2 μm, and the minimum distance between PFCs of adjacent submodules would be 11-12 μm, corresponding to a loss of one PFC in each row and column of 10 μm [1, 9, 10, 28, 37-41, 43, 44].

In the case of CRT photodiodes, the minimum distance between the edge PFC s of adjacent submodules will be 17-19 μm (with a gap between the crystals of adjacent submodules of not more than 3 μm), which corresponds to the loss of two PFCs in each row and column [1, 9, 10, 27, 28, 37-41, 43, 44]. A more complete volume of visualized information in the subsequent processing of photographic signals improves the quality of the generated images.

Application of technological methods, including optical methods, in which the minimum distance between the edge PFCs of adjacent submodules will not exceed 2-3 microns, will ensure the absence of element losses in each row and column. "Blind zones" will be optically overlapped by adjacent PFCs.

In this case the signal reading is performed without loss of video information in each frame, which provides the maximum (*100%*) efficiency of image conversion in MFP [1, 9, 10, 37-41, 43, 44].

10. CONCLUSION

The fundamental results of the conducted research into the technology of creating MFPs of ultra-high dimensionality with maximum efficiency of conversion of multidimensional signals, including the development of an optical methodology for achieving the ultimate efficiency of image conversion, are as follows. Microminiaturization of the MFP technological design provides an increase in the percentage of good devices and the corresponding breakthrough expansion of the MFP industrial sphere of application in advanced IT-productions and promising scientific research, preservation of the given FFE step in the field of crystal microdocking and, in some cases, complete elimination of losses of visualized information in MFP, as well as increasing the MFP format to extra-high dimensionality and more.MFPs of extra-high dimension with maximum efficiency of image transformation provide high clarity of formed images, high productivity of parallel reading of video data, increase of sensitivity and significant increase of spatial resolution and speed of receiving systems in IR and THz ranges. Special super high-dimensional MFPs are applicable for particle registration in a wide range of energies [107, 108]. Fundamental approaches in the study of the mosaic technology of creating ultrahigh-dimensional FPs provide continuous monitoring of large data, distributed processing of information arrays, multidimensional signals and sequences of ultrahigh-format images in different spectral ranges, in real time.The author is grateful to Academician RAS A. V. Latyshev for support of research; Prof., Dr. Phys. and Math. N. Ovsyuk for fruitful discussion of the results of fundamental   research4;   M. V. Yakushev, Ph. B. M. Iakushev, Ph.D., for their help in the development of the methodology for the creation of MFPs based on various semiconductor materials4; M.A. Vasiliev, Ph.M.A. Demianenko for numerous discussions on microbolometers   and   ;   Associate Professor A.G. Kharlamov for consideration of the extended application area of MFPs and useful discussions of fundamental approaches in the research of creating MFPs of high and ultra-high dimensions.

LITERATURE

1. Demianenko M. A., Kozlov A. I., Novoselov A. R., Ovsyuk V. N. Increasing the efficiency of image conversion in mosaic microbolometer receivers // Optical Journal. 2018. T. 85. № 2. C. 60–66.

2. Kozlov A. I. Analysis of the principles of silicon multiplexer circuits for multielement IR photodetectors // Autometry. 2010. T. 46. № 1. C. 118–129.

3. Kozlov A. I. Design Features and Some Realizations of Silicon Multiplexers for Infrared Photodetectors // Optical Journal. 2010. T. 77. № 7. C. 19-29.

4. Vasiliev V. V., Kozlov A. I., Marchishin I. V., Sidorov Yu. G., Yakushev M. V. Analysis of structural and technological limitations in the signal reading circuits of infrared photodiodes // Optical Journal. 2014. T. 81. № 7. C. 39-45.

5. Demianenko M. A., Kozlov A. I., Marchishin I. V., Esaev D. G., Ovsyuk V. N. Investigation of technological limitations in silicon signal reading circuits of infrared photodetectors based on multilayer structures with quantum wells // Avtometriya. 2015. T. 51. № 2. C. 110-118.

6. Kozlov A. I., Demianenko M. A., Ovsyuk V. N. Optimization of parameters of the system "infrared photosensitive element based on multilayer structures with quantum wells - silicon multiplexer of photographic signals // Optical Journal. 2017. T. 84. № 9. C. 59-65.

7. Kozlov A. I., Demianenko M. A., Marchishin I. V., Ovsyuk V. N. Creation of analog-digital silicon multiplexers of infrared photodetector signals // Autometry. 2016. T. 52. № 6.

C. 120–127.

8. Kozlov A. I., Demianenko M. A., Ovsyuk V. N. Analytical comparison of the characteristics of infrared photodetectors based on HgCdTe photodiodes and GaAs/AlGaAs photodetectors with quantum wells // Optical Journal. 2016. T. 83. № 9. C. 64-71.

9. Kozlov A. I., Demianenko M. A., Novoselov A. R., Ovsyuk V. N. Application of equivalent noise temperature difference for comparison of ultrahigh-dimensional photodetectors based on multilayer structures with quantum wells // Optical Journal. 2020. T. 87. № 1. C. 37–44. DOI: 10.17586/1023-5086-2020-87-01-37-44.

10. Kozlov A. I., Novoselov A. R., Demianenko M. A., Ovsyuk V. N. Mosaic

photodetector with the maximum image conversion efficiency: structures and manufacturing methods (variants) // Russian patent number RU2731460C1. Publ. 03.09.2020. Bulletin No. 25.

11. Gulbransen D. J., Love P. J., Murray M. P., Lum N. A., Fletcher C. L., Corrales E., Mills R. E., Hoffman A. W., Ando K. J. Megapixel and Larger Readouts and FPAs for Visible and Infrared Astronomy // Proc. of SPIE. 2003. V. 4841. P. 770-781.

12. Kozlowski L., Bay Y., Vural G., Vural K. Low-Noise Monolitic and Hybrid CMOS-based Sensors // Rockwell Scientific. Nov. 2001. P. 1-26. URL: http://www.noao.edu/meetings/lsst/kozlowski.pdf (accessed 04.09.2020).

13. Dorn R. J., Finger G., Huster G. et al. The CRIRES Megapixel Focal Plane Array Detector Mosaic // Proc. of SPIE. 2004. V. 5499. P. 510–517.

14. Overclockers.ru: RussOver. Technological processes of semiconductor processor manufacturing // URL: http://overclockers.ru/russover/show/13-984/technologicheskie_processy_poluprovodnikovogo_proizvodstva_processorov (accessed 05.09.2020).

15. Gotra Z. Yu. Technology of microelectronic devices // M.: Radio and Communications. 1991. 528 c.

16. Aseev A. L. Nanomaterials and nanotechnologies for modern semiconductor electronics // Russian Nanotechnologies. Nanoobzor. 2006. № 6. C. 97–110.

17. Rogalski A. Progress in Focal Plane Array Technologies // Progress in Quantum Electronics. 2012. № 36. P. 342–473.

18. Agnese P., Cigna C., Pornin J.-L., Accomo R., Bonnin C., Colombel N., Delcourt M., Doumayrou E., Le Pennec J., Martignac J., Reveret V., Rodriguez L., Vigroux L. G. Filled Bolometer Arrays for Herschel / Pacs // Proc. of SPIE: Millimeter and Submillimeter Detectors for Astronomy. 2003. V. 4855. P. 108–114. DOI: 10.1117/12.459191.

19. Harwit M. , Helou G., Armus L., Bradford C. V. et al. Far- Infrared / Submillimeter Astronomy from Space // Workshop: Recommendations for The Astronomy and Astrophysics Decadal Survey of 2010. CA. Jan. 2009. P. 1-39. URL: http://www.asd.gsfc.nasa.gov/cosmology/spirit/fir-sim_crosscutting_white_pa- per.pdf (дата обращения: 04.09.2020).

20. Hall D. N. B., Luppino G., Hodapp K. W., Garnett J. D., Loose M., Zandian M. A 4096×4096 HgCdTe Astronomical Camera enabled by the JWST NIR Detector Development Program // Proc. of SPIE. V. 5499. Optical and Infrared Detectors for Astronomy. Sep. 2004. P. 1-15. DOI: 10.1117/12.554733.

21. Rogalski A. Next decade in infrared detectors // Proc. of SPIE. Electro- Optical and Infrared Systems: Technology and Applications XIV. Warsaw, Poland. Oct. 2017. V. 10433. P. 104330L-1 - 104330L-25. DOI: 10.1117/12.2300779.

22. Reveret V., Andre Ph., Talvard M., Rodriguez L. R., Boulade O., Doumayrou E., Gallais P., Horeau B., Le Pennec J., Lortholary M., Martignac J., Agnese P. Herschel / Pacs Bolometer Arrays for Submillimeter Ground-based

Telescopes // Journal of Low Temperature Physics. 2008. V. 151. P. 32-39. DOI: 10.1007/s10909-007-9705-2.

23. Finger G., Beletic J. W. Review of the state of infrared detectors for astronomy in retrospect of the June 2002 Workshop on Scientific Detectors for Astronomy // Proc. of SPIE. 2003. V. 4841. P. 839-852.

24. Rossi, L., Fischer, P., Rohe, T., Wermes, N. Pixel Detectors. From Fundamentals to Applications // Springer. 2006. 304 p. URL: http://www.springer.com/gp/book/9783540283324 (accessed 04.09.2020).

25. Yakovleva N. I., Boltar K. O., Nikonov A. V. Multicrystal multicolor photodetector with extended spectral quantum efficiency characteristic // Patent No. RU2564813C1. Published: 10.10.2015. Bulletin No. 28.

26. Chamonal J. P., Mottin E., Audebert P., Ravetto M., Caes M., Chatard J. P. Long linear MWIR and LWIR HgCdTe arrays for high resolution imaging // Proc. of SPIE. 2000. V. 4130. P. 452-462.

27. Novoselov A. R. Development of highly efficient mosaic photodetectors based on lines of photosensitive elements // Autometry. 2010. T. 46. № 6. C. 106–115.

28. Novoselov A. R. Method of chip facet formation for mosaic photodetector modules // Patent № RU2509391C1. Published: 10.03.2014. Bul. № 7.

29. Liddiard K. C. Extending the Reach of Mosaic Pixel Infrared Focal-Plane Arrays // Proc. of SPIE: IR Technology and Applications XXXIX. 2013. V. 8704. P. 8704-

3F. DOI: 10.1117/2.1201110.003904.

30. Demianenko M. A., Kozlov A. I., Novoselov A. R., Klimenko A. G., Marchishin I. V., Esaev D. G., Ovsyuk V. N. Research of technology of creation of mosaic uncooled infrared microbolometer receivers

and terahertz spectral ranges up to 3072×576 and more // Proceedings of the First Russian-Belarusian scientific-technical conference "Elemental base of domestic radioelectronics", September 11-14, 2013. Nizhny Novgorod: A.S. Popov Scientific and Technical Society for Radio Engineering, Electronics and Communications. A. S. Popov. 2013. C.30-33.

31. Demianenko M. A., Kozlov A. I., Novoselov A. R., Klimenko A. G., Marchishin I. V., Esaev D. G., Ovsiuk V. N. Research of constructive and technological ways of creating uncooled microbolometer image receivers of infrared and terahertz spectral ranges of up to 3072×576 and more // Abstracts of XI Russian Conference on Semiconductor Physics, "Semiconductors-2013", 16-20 September 2 01 3 St. Petersburg: Ioffe Physical-Technical Institute. A. F. Ioffe RAS. 2013. C. 43.

32. Demianenko M. A., Kozlov A. I., Novoselov A. R., Klimenko A. G., Marchishin I. V., Esaev D. G., Ovsyuk V. N. Image conversion in mosaic uncooled microbolometer receivers of infrared and terahertz ranges with format up to 3072×576 and more // Optical Journal. 2014. T. 81. № 3. C.35-43.

33. Demianenko M. A., Kozlov A. I., Novoselov A. R., Klimenko A. G., Marchishin I. V., Esaev D. G., Ovsyuk V. N. Development of mosaic uncooled microbolometer receivers of infrared and terahertz spectral ranges of format up to 3072×576 and more // Advances in Applied Physics. 2014. T. 2. № 2. C.123-130.

34. Kozlov A. I., Novoselov A. R., Demyanenko M. A., Esaev D. G., Ovsyuk V. N. Principle Development Mosaic to Create Panoramic with Cooling Microbolometer (Overview) Infrared Image Detector and Terahertz Spectral Range // Italian Science Review. 2015. V. 28. № 7. P. 11–15.

35. Demyanenko M.A., Kozlov A.I., Novoselov A.R., Esaev D.G., Ovsyuk V. N. Development of mosaic principle of uncooled large-format microbolometer receivers // Proceedings of the XII International Conference "Applied Optics - 2016". November 14-18, 2016. St. Petersburg: D.S. Rozhdestvensky Optical Society. D. S. Rozhdestvensky. 2016. In 3 volumes. T. 2. C. 77–82.

36. Demianenko M.A., Kozlov A.I. , Novoselov A. R. Ovsyuk V. N. On mosaic uncooled microbolometer receivers of infrared and terahertz ranges // Autometry. 2016. T. 52. № 2. C.115-121.

37. Novoselov A. R. A method for reducing the chip gap in mosaic photodetector modules // Autometry. 2016. T. 52. № 1. C. 116–121.

38. Demianenko M. A., Kozlov A. I., Novoselov A. R., Esaev D. G., Ovsyuk V. N. Mosaic principle of uncooled microbolometer receivers of ultrahigh dimensionality // Theses of reports of the Russian conference and school of young scientists on topical problems of semiconductor photoelectronics, "Photonics-2017". Novosibirsk: IPC NSU, 2017. C. 153.

39. Demianenko M. A., Kozlov A. I., Novoselov A. R., Ovsyuk V. N. Comparative analysis of the mosaic principle of creation and characteristics of high-dimensional infrared photodetectors based on multilayer structures with quantum wells // Proceedings of XXV International Scientific and Technical Conference and School on Photoelectronics and Night Vision Devices, May 24-26, 2018, Moscow. In 2 volumes. - M.: JSC "Orion", M.: Publishing house "OFSET MOSCOW", 2018. T. 2. C. 530–533.

40. Demianenko M. A., Kozlov A. I., Novoselov A. R., Ovsyuk V. N. Improvement of Image Conversion Efficiency in Mosaic

microbolometer receivers // Proceedings of the XXV International Scientific and Technical Conference and School on Photoelectronics and Night Vision Devices, May 24-26, 2018, Moscow. In 2 volumes. - M.: JSC "Orion", M.: Publishing house "OFSET MOSCOW", 2018. T. 2. C. 357–360.

41. Kozlov A. I., Novoselov A. R. Technological approaches to the creation of mosaic photodetectors of ultra-high dimension with the maximum image conversion efficiency // Abstracts of the Russian conference on current problems of semiconductor photoelectronics, "Photonics-2019", Novosibirsk. May 27-31 , 2019 P . 152. DOI: 10/34077/RCSP2019-152.

42. Kozlov A. I., Novoselov A. R., Ovsyuk V. N. Design principles of image loss elimination in mosaic photodetectors of ultrahigh dimensionality // Abstracts of the Russian Conference on Current Problems of Semiconductor Photoelectronics, Photonics-2019", Novosibirsk. May 27-31, 2019 P. 171. DOI: 10/34077/RCSP2019-171.

43. Kozlov A. I. , Novoselov A. R. , Demianenko M. A. , Ovsyuk V. N. On mosaic infrared photodetectors of ultra-high dimension based on multilayer structures with quantum wells // Abstracts of the Russian Conference on Current Problems of Semiconductor Photoelectronics, "Photonics-2019", Novosibirsk. May 27-31, 2019 P. 140. DOI: 10/34077/RCSP2019-140.

44. Kozlov A. I., Novoselov A. R., Demianenko M. A., Ovsyuk V. N. Fundamental bases of creating mosaic photodetectors of ultrahigh dimensionality with limiting image conversion efficiency // Theses of reports of XIV Russian conference on semiconductor physics, "POLUPROVODNIKI-2019", Novosibirsk. September 9-13, 2019. In 2 vols. T. 2. C. 447. DOI: 10/34077/Semicond2019-447.

45. Kozlov A. I. Features of the influence of ΔN and $\Delta \mu$ *1/f-noise* models on the noise equivalent temperature difference of infrared photodetectors // Abstracts of the XIV Russian Conference on Semiconductor Physics, "POLUPROVODNIKI-2019", Novosibirsk. September 9-13, 2019. In 2 volumes. T. 2. C. 446. DOI: 10/34077/Semicond2019-446.

46. Kozlov A. I. Conceptual congruence of photodetectors based on superlattices and multilayer structures with quantum wells // Abstracts of XIV Russian Conference on Semiconductor Physics, "POLUPROVODNIKI-2019", Novosibirsk. September 9-13, 2019. In 2 volumes. T. 2. C. 445. DOI: 10/34077/Semicond2019-445.

47. Broennimann Ch., Glaus F., Gobrecht J., Heising S., Horisberger M., Horisberger R., Kastli H.C., Lehmann J., Rohe T., Streuli S. Development of an Indium-Bump Bond Process for Silicon Pixel Detectors at PSI // Nuclear Instruments and Methods in Physics Research Section A: Accelerators, Spectrometers, Detectors and Associated Equipment. 2006. V. 565. № 1. P. 303–308.

48. Bortoletto F., Bonoli C., D'Alessandro M., Giro E., De Caprio V., Corcione L., Ligori S., Morgante G. Euclid Near Infrared Spectrograph (ENIS) Focal Plane Design // Proc. of SPIE. V. 7731. Space Telescopes and Instrumentation 2010: Optical, Infrared, and Millimeter Wave. Aug. 2010. P. 7731-2T. DOI: 10.1117/12.856402.

49. Sprafke T., Beletic J. W. High-Performance Infrared Focal Plane Arrays for Space Applications // Optics and Photonics News. 2008. V. 19. № 6. P. 22–27. Figure in center of P. 27. DOI: 10.1364/OPN.19.6.000022.

50. Klimenko A. G., Novoselov A. R., Ovsyuk V. N. et al. Technology of assembly of large-format infrared photodetector modules on indium microtubes // Optical Journal. 2009. Т. 76. № 12. С. 63-68.

51. Miller T. M., Jhabvala Ch. A., Leong Ed., Costen N. P., Sharp E., Adachi T., Benford D. J. Enabling Large Focal Plane Arrays Through Mosaic Hybridization // Proc. of SPIE. High Energy, Optical, and Infrared Detectors for Astronomy, Part V. 2012 . Vol. 8453. P. 2H-1-2H-7. See Fig.5. DOI: 10.1117/12.926491.

52. Kozlov A. I. Research on creation of mosaic photodetectors with limiting efficiency of ultra-high dimensional image conversion in optical, infrared and terahertz ranges. Conceptual review. - Novosibirsk: Novosibirsk Publishing House, 2020. - 168 p., ill. ISBN 978-5-6045277-2-6.

53. Kozlov A. I. Congruence analysis of ultra-high dimensional photodetectors with maximum image conversion efficiency based on multilayer structures with quantum wells. Analytical review. - Novosibirsk: A. I. Kozlov Publisher, 2020. - 86 p., ill., ISBN 978-5-6045564-1-2.

54. Kozlov A. I. Research on creation of mosaic photodetectors with ultimate efficiency of high and ultra-high dimensional image conversion in optical, infrared and terahertz ranges. Conceptual Review of Variants. Vol. 2, revised. - Novosibirsk: A.I. Kozlov Publisher, 2020. - 133 p., ill., ISBN 978-5-6045564-0- 5.

55. Rotolante R. A., Koehler T. Focal Plane Photo-detector Mosaic Array Fabrication // Patent No. US4290844A. Publ. 22/09/1981.

56. Rotolante R. A., Koehler T. Focal Plane Photo-detector Mosaic Array Apparatus // Patent No. US4449044A. Publ. 15.05.1984.

57. Carson J. C., Clark C. A., Thermal Imager Incorporating Electronics Module Having Focal Plane Sensor Mosaic // Патент № US4806761A. Опубл. 21.02.1989.

58. Liddiard K. C. Microbolometer Infrared Security Sensor // Application for Invention patent No. NZ560809A. Published on 25.09.2009.

59. Bachs M. B., Casas E. C., Fantoba M. L., Graells F. S., Teres L. T. Modular Hybrid Device For Reading Image Sensor Arrays // Patent application for invention number WO2007006836A1. Published 18.01.2007.

60. Miyake M., Boyd W. E. High Density Array Module and Connector // Application for an invention patent No. US20110147568A1. Published on June 23, 2011.

61. Alman D. H., Griffus D. L., Canning R. V., Slayton (Jr.) M. C. Process for DisplayingandDesigningColors//. Application for   patent application   for invention
NO. WO2013063552A1. Published on May 2, 2013.

62. Bochkov V. D., Drazhnikov B. N. Photoreceiving device. // Patent. NO. RU2392691C1. Published: June 20, 2010. Bulletin no. 17.

63. Murayama D. Image formation device, image processing method and integrated circuit // Patent № RU2367109C2. Publ. 10.09.2009. Bulletin number 25.

64. Fukuhara T. Image capture device, integrated circuit of image capture element and method of image capture result processing // Patent No. RU2367108C2. Publ. 10.09.2009. Bulletin number 25.

65. Fukuhara T., Narabu T. Image formation device, integrated circuit for image capture device and image data processing method // Patent No. RU2367107C2. Publ. 10.09.2009. Bulletin number 25.

66. Netesin N. N., Korotkova G.P., Korzenev G. N., Povolotsky S. N., Karpova N. V., Korolev O. V., Baranov R. V., Povolotsky G. Yu. Method of manufacturing thin-film multilevel boards for multichip modules, hybrid integrated circuits and microcircuits // Patent
NO. RU2459314C1. Published: 20.08.2012. Bulletin No. 23.

67. Szilveszter J. Microbolometer Focal Plane Array with Integrated Multi-Spectral Mosaic Band-Pass Filter/Focusing Lens Array for Simultaneous Real-Time

Anesthetic and Respiratory Gas Concentration Detection and Measurement Signal Processing // Патент № CN110088595A. Опубл. 02.08.2019.

68. Xiuping Zh. Waste Gas Treatment Device for Glass Mosaic Production Device // Патент № CN109316926A. Опубл. 12.02.2019.

69. Jianbo G., Shuyan Zh., Xia Zh., Qian H., Zhifeng G., Chuanfen L. Test Method and Test Device for Detecting Mosaic Angle Distribution of Neutron Monochromator // Патент № CN109212585A. Опубл. 15.01.2019.

70. Pichette J., Task N. Hyperspectral Imager for Snapshot and Line-scan Modes //

Patent No. US2018359438A1. Republished 12/13/2018.

71. Piaoyang Ch. Novel Glass Mosaic's Exhaust Purification Device // Патент NO. CN208115499U. Published November 20, 2018.

72. Sheng L. Gas Detector and Gas Controller Data Transmission's Device // Patent No. CN207557195U. Published on 29.06.2018.

73. Weilu Ya. Portable How Gaseous Detector // Patent No. CN207352753U. Published May 11, 2018.

74. Dixon A. E. Scanning Microscope Using A Mosaic Scan Filter // Патент NO. CA3035109A1. Published March 01, 2018.

75. Xingbo W., Qirong L. Automatic Sticking Device for Mosaic Stickers // Patent No. CN104553593A. Published 30.06.2017.

76. He Ye., Xingbo W., Fei Zh. Many Colors Mosaic Grain Sorting Mechanism Based on Visual Identification // Патент № CN206104397U. Опубл. 19.04.2017.

77. Yetzbacher M. K., Wilson M Multiple Band Short Wave Infrared Mosaic Array Filter // Invention Application No. WO2016040755A1. Published on 09.02.2017.

78. Xingbo W., He Ye., Fei Zh. Multi-Color Mosaic Particle Classifying Mechanism Based on Visual Recognition // Патент № CN106269578A. Опубл. 04.01.2017.

79. Yueming W., Shiyao Zh., Jianyu W., Rong Sh., Junwei L., Xizhong X., Liyin Y., Wenjun H., Zhikang B. Object Space View Field Mosaic Infrared Hyper- Spectral Imaging System // Patent No. CN104535182A. Published on 07.12.2016.

80. Bowler D. P., Wein S. J., Korwan D. J., Targove J. D., Perron G. M. Multi Field of View Imaging System // Patent No. US9325883A1. Published April 26, 2016.

81. Bingquan L., Zhang J., Wen Q. Radiograph Mosaic Device and Method // Patent No. CN102551742A. Published on September 23, 2015.

82. Xuguo Zh., Zhaorong L., Dakai Zh., Jingjing G., Nan Zh., LI W., Yuedong Zh. Lens Light Splitting Mode-Based Focal Plane Splicing Aerial Surveying Camera with Extra Large Plane // Патент № CN103143839A. Опубл. 17.12.2014.

83. Justice J. Multi-Spectral Sensor System Comprising a Plurality of Buttable Focal Plane Arrays // Invention Patent Application No. US20140110585A1. Published

April 24, 2014.

84. Bartholomew J. L. Filter Mosaic for Detection of Fugitive Emissions // Patent No. US8559721B1. Published on 15.10.2013.

85. Yong L., Zhangcheng Ya., Shenghui L., Xuepeng W., Anbing G., Tao X., Quan Zh. Infrared Mosaic Imaging Device, Patent No. CN101685203A. Published on 08.08.2012.

86. Sultan M. F., Harbach A. P., Kuhlman F. F. Seamless Mosaic Projection System and Method of Aligning the Same // Patent application for invention NO. US20120169684A1. Published on July 05, 2012.

87. Yafuso E. System for Mosaic Image Acquisition // Application for patent number US20110115916A1. Published on May 19, 2011.

88. Baolong G., Zhanlong Ya., Yunyi Y. Image Mosaic Method Based on Neighborhood Zernike Pseudo-Matrix of Characteristic Points // Патент NO. CN101556692A. Published on 10/14/2009.

89. Wilson D. W., Bearman G. H., Johnson W. R. Color Camera Computed Tomography Imaging Spectrometer for Improved Spatial-Spectral Image Accuracy // Patent application for invention No. WO2007070610A2. Published on 10.07.2008.

90. Clark S. A. Detector Array Focal Plane Configuration// Patent NO. US4403238A. Published 06/09/1983.

91. Lorenze (Jr.) R. V., White W. J. Hybrid Mosaic IR/CCD Focal Plane //. Patent No. US4286278A. Published on 08/25/1981.

92. Lorenze (Jr.) R. V. DurableInsulatingProtectiveLayerfor Hybrid CCD/Mosaic IR Detector Array," Patent No. US4275407A. Published June 23, 1981.

93. Lorenze (Jr.) R. V., White W. J. Hybrid Mosaic IR/CCD Focal Plane //. Patent No. US4197633A. Published April 15, 1980.

94. Lorenze (Jr.) R. V. Durable Insulating Protective Layer for Hybrid CCD/Mosaic IR Detector Array // Патент № US4196508A. Опубл. 08.04.1980.

95. Brown K. A. Orthoscopic Image Tube // Patent No. US3878329A. Published April 15, 1975.

96. Andreev V. M., Ilyinskaya N. D., Malevskaya A. V., Rumyantsev V. D. Solar

photovoltaic submodule // Patent No. RU2442244C1. Published on February 10, 2012. Bulletin number 4.

97. Kozlowski L. J., Bailey R. B., Cabelli S. C., Cooper D. E., McComas G., Vural K., Tennant W. E. 640×480 PACE HgCdTe FPA // Proc. of SPIE. 1992. V. 1735. P. 163-174.

98. DeWames R. E., Arias J. M., Kozlowski L. J., Williams G. M. An assessment of HgCdTe and GaAs/GaAlAs technologies for LWIR infrared imagers // Proc. of SPIE. 1992. V. 1735. P. 2-16.

99. Demianenko M. A., Kozlov A. I., Ovsyuk V. N. et al. Shift register (variants) // Russian patent number RU2530271C1. Published on 10.10.2014. Bulletin № 28.

100. Kozlov A. I., Marchishin I. V., Ovsyuk V. N., Shashkin V. V. Silicon multiplexers for multielement IR photodetectors

range // Autometry. 2005. T. 41. № 3. C. 88-99.

101. Kozlov A. I., Marchishin I. V., Ovsyuk V. N. Silicon multiplexers 320×256 for infrared photoreceivers on the basis of CRT-diodes (in Russian) // Avtometriya. 2007. T. 43. № 4. C. 74-82.

102. Kozlov A. I., Marchishin I. V. Industrial-oriented development of silicon multiplexers for multielement IR photodetectors // Autometry. 2012. T. 48. № 4. C. 60-72.

103. Demianenko M. A., Kozlov A. I., Klimenko A. G., Marchishin I. V., Ovsyuk V. N., Savchenko A. P., Toropov A. I., Shashkin V. V. Matrix IR-photodetectors based on multilayer GaAs/AlGaAs heterostructures with quantum wells // Applied Physics. 2000. № 6. C. 94-100.

104. Esaev D.G., Marchishin I. V., Ovsyuk V. N., Savchenko A. P., Fateyev V. A., Shashkin V. V., Sukharev A. V., Padalitsa A. A., Budkin I. V., Marmalyuk A. A. Infrared photodetector on the basis of GaAs/AlGaAs multilayer structures with quantum wells // Autometry. 2007. T. 43. № 4. C. 112-118.

105. Kozlov A. I., Marchishin I. V., Ovsyuk V. N., Aseev A.L. Series of silicon multiplexers for CRT - photodiodes of 8-16 μm spectral range // Optical Journal. 2008. T. 75. № 3. C. 60-67.

106. SOFRADIR products: Refrigerated IR FPs // Official website of SOFRADIR Group. URL: http://www.sofradir- ec.com/products-cooled.asp.

107. Anashin V. V., Aulchenko V. M., Baldin E. M. and KEDR Collaboration. Measurement of $\Gamma ee(J/\psi)$ with KEDR detector // J. High Energ. Phys. 2018. № 5. P. 119–139.

108. Aad G., Abbott D., Abdallah J. and ATLAS Collaboration. Search for heavy ZZ resonances in the $l+l-l+l-$ and $l+l-vv-$ final states using proton-proton collisions at $\sqrt{s}-= 13$ TeV with the ATLAS detector // Eur. Phys. J. C. 2018. V. 78. P. 293–326.

109. Kozlov A. I., Raigel V. I., Spector A. A. Study of qualitative characteristics of two-dimensional vector filter // Theses of reports of the scientific conference of young scientists devoted to the memory of Academician L. V. Kirensky. - Krasnoyarsk. 8-13 apr. 1985. p. 155.

Alexander I. Kozlov was born in Siberia, Russia, in 1962. From 1984 to the present time, he is with ISP SB RAS. He received the PhD in technical science in 1995. From 2001, he is the Senior Researcher.

He is the author of the 9 books, the 93 articles, and the 14 inventions. His research interests include micro- and nanophotoelectronics, semiconductor-structure physics, model building, schematics of ICs, chip design, IR and THz FPAs, mosaic FPAs, microbolometer, MCT, QWIPs, ROICs.
ORCID: 0000-0001-9806-5985.

phone·
s.t. +7-913-762-65-90; d.t. +7-383-265-37-82.

Candidate of Technical Sciences Kozlov Alexander Ivanovich.

01.01.2021 г.

UNDAMENTAL BASES FOR CREATING ULTRA-HIGH DIMENSIONAL MOSAIC PHOTODETECTORS WITH MAXIMUM IMAGE CONVERSION EFFICIENCY

Alexander I. Kozlov

CONTENTS

Retro Page [109]:

ИССЛЕДОВАНИЕ КАЧЕСТВЕННЫХ
ХАРАКТЕРИСТИК ДВУМЕРНОГО ВЕКТОРНОГО ФИЛЬТРА

Козлов А.И.,Райгель В.И.,Спектор А.А.
630090,Новосибирск,Институт физики полупроводников СОАН СССР.

В работе исследованы качественные показатели квазиоптимального рекурсивного фильтра калмановской структуры , осуществляющего двумерную обработку реальных изображений с целью восстановления визуальной информации. Подобная обработка необходима , например , в случаях искажения изображения вследствии его смаза или расфокусировки оптики. Для описания искажённых изображений использована двумерная векторная марковская модель второго порядка /I/.

Оценка качественных характеристик исследуемого фильтра произведена по среднеквадратической ошибке восстановления и визуальному восприятию восстановленных изображений. Для сравнения рассчитана ошибка оптимального фильтра. Проведён эксперимент по восстановлению изображений с искажениями , смоделированными с помощью ЭВМ.

Выполненные расчёты и эксперименты показывают , что ошибка исследуемого фильтра в несколько раз больше ошибки оптимального фильтра , при этом однако качество визуального восприятия восстановленного изображения довольно высокое.

Для увеличения эффективности работы фильтра необходимо переходить к описанию изображений моделями более высокого порядка.

Литература:

I. Методы цифровой обработки изображений : Межвуз.сб. науч.трудов / Новосиб.электротехн.ин-т ; Отв.ред. Т.Б.Борукаев . - Новосибирск , 1983 . - II7 с .

Summary:

Expanded title of the analytical review:

"FROM FUNDAMENTALS TO ULTRA-HIGH-DIMENSIONAL MOSAIC PHOTODETECTOR TECHNOLOGY WITH ULTIMATE IMAGE CONVERSION EFFICIENCY"

Purpose of this review: **STUDYING THE PRIMARY NANO- AND MICROMINIATURIZATION OF MOSAIC PHOTOPREMISTS OF HIGH AND ULTRAORDINARY SIZES**

The following publications of 2020 are available on this topic: a patent of the Russian Federation for an invention, an article and monographs by A. I. Kozlov:

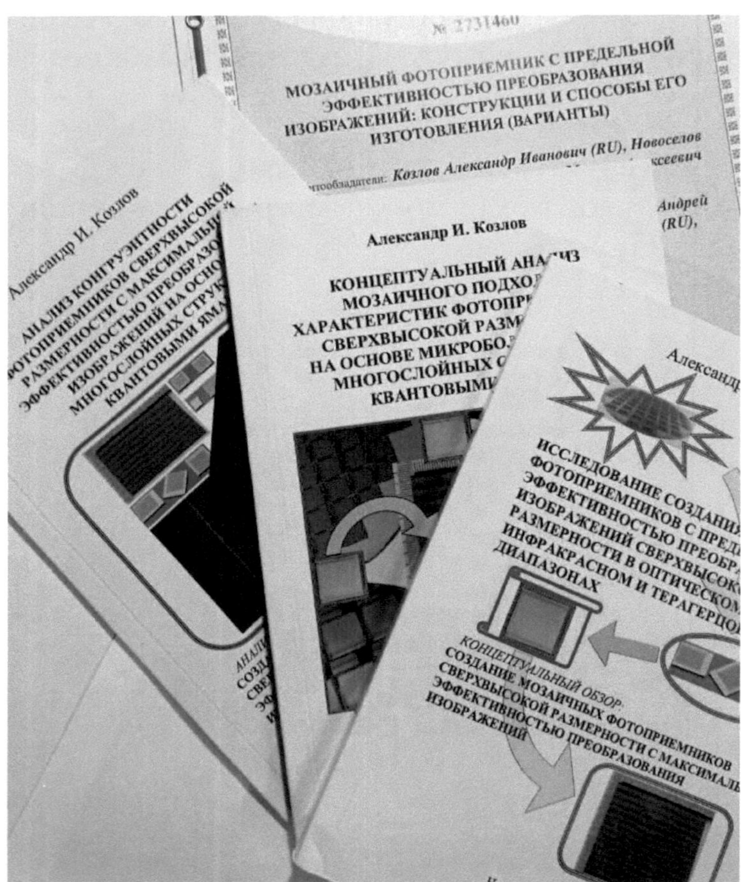

FUNDAMENTALS OF THE CREATION OF ULTRA-HIGH-DIMENSIONAL MOSAIC PHOTODETECTORS WITH MAXIMUM IMAGE CONVERSION EFFICIENCY

Kozlov Alexander Ivanovich, Ph.

ANALYTICAL REVIEW: STUDY OF THE ULTIMATE NANO- AND MICROMINIATURIZATION OF ULTRA-HIGH DIMENSIONAL MOSAIC PHOTODETECTORS

Prepared for print by Kozlov Alexander Ivanovich (author). Address: 630084, Russia, Novosibirsk, 12 Respublikanskaya St., kv. 24; Kozlov Alexander Ivanovich; *e-mail:* aikozlov13@mail.ru

Phone numbers: s.t. +7-913-762-65-90;

d.t. 8 (383) 265-37-82 (*you can leave a message on the answering machine*).

Responsible for the issue - A.I. Kozlov,
e-mail: aikozlov13@mail.ru .
The technical editor of the issue is V. N. Rudenko,
e-mail: r.valentina@lap-publishing.ru .

Издательство "Lambert Academic Publishing" ("LAP"); Brivibas gatve 197, Riga, LV-1039, Latvia, European Union; Reg. No. 90012088472, *web-site:* **https://www.lap-publishing.com/**

Signed into print *on January 13, 2021*. Order №/2021____ .
Format 60x84/16. Digital printing. Circulation__________copy.
The volume is 5.00 typewritten pages. _,__ p. l. Paper format A5.
Page 82. Table 2. Ill. 39. Bibliography. 109 titles.

Printed by Books on Demand GmbH, Norderstedt / Germany